Economic Entomology
in the Tropics

Economic Entomology in the Tropics

K. P. LAMB

Department of Biology,
University of Papua New Guinea,
Port Moresby, Papua New Guinea

1974

Academic Press   London · New York

A Subsidiary of Harcourt Brace Jovanovich, Publishers

ACADEMIC PRESS INC. (LONDON) LTD
24–28 Oval Road,
London NW1

U.S. Edition published by
ACADEMIC PRESS INC.
111 Fifth Avenue,
New York, New York 10003

Library of Congress Catalog Card Number: 73–19017

ISBN: 0–12–434650–2

Text set in 10/12 pt. Monotype Times New Roman, printed by letterpress
in Great Britain at The Pitman Press, Bath

Preface

This book is orientated towards the needs of developing countries in the humid tropics and emphasis is given to the agricultural pests of the Oriental and Australian regions. However, the major pests of crops from other parts of the tropics are also considered as are insects of medical and veterinary importance.

It is assumed that the reader has some background experience of general zoology or basic entomology. There are many books dealing with the classification of insects and their physiology or ecology and some books dealing with the pests of specific crops but few attempts have been made to bring together the large but scattered body of information on tropical insect pests. It is hoped that this book will provide an introduction to the subject and further reading is provided at the end of each chapter.

The densest populations of man are found in the tropics and it is in this third world that pests and diseases take their greatest toll of the crops and health of man and domestic animals. In an increasingly hungry world this can no longer be tolerated. With the passing of colonialism there has been a decline in scientific effort relating to tropical diseases and pests. This trend must be reversed by increased education and investigation within the developing countries themselves. That is the rationale for this book.

K. P. LAMB
Papua New Guinea   *September* 1973

Contents

Chapter 1   Insects, Good and Bad

From the evolutionary point of view, insects are the most successful group of land animals since they colonize every conceivable terrestrial habitat and the number of species known greatly exceeds that of any other order. The abundance of insects is brought home particularly forcefully in the tropics where higher temperatures and humidities often lead to large populations of spectacular, annoying or dangerous species. The insects compete successfully with man for his food; they attack his clothing and houses, his livestock and his pets. In many places they carry diseases which may even make some areas unliveable for man or his domestic animals.

The majority of insects are neither directly harmful nor obviously beneficial to man and we tend to forget our great debt to the insects for their roles in plant pollination, in terrestrial food chains and in control of their own kind—quite apart from the relatively minor contributions they have made towards feeding or clothing man. They have also proved to be very suitable experimental animals for biological investigation and their short life cycles and ready adaptability to life in the laboratory have been exploited for experimental work in genetics, ecology, physiology and biochemistry.

One of the best known insect products which has long been used commercially is silk produced from the salivary glands of Saturniid moths. The silkworm *Bombyx mori* (L.) has been cultivated in China for 5,000 years. The silk trade from China to the Mediterranean was established about the beginning of the first century BC. Silk caravans passed between China and Iran at least as far back as 106 BC. Silkworms were introduced into Japan about 300 AD and into India a little later and these three countries have remained major silk producers. In 552 AD the silkworm (and mulberry seeds) were introduced into Europe and 10 years later the silk industry was declared a state monopoly by the Roman emperor Justinian. Today the greatest quantity of silk is produced in Asia where labour costs are low (Table 1). World production in 1965 was 33,000 metric tons, worth about 753 million US dollars.

Despite severe competition from cheaper synthetic materials silk has some highly desirable qualities such as high, dry tensile strength, toughness, high

Table 1.
World production of silk in 1965 (in metric tons)

Japan	19,106
China	7,500
India	1,284
South Korea	851
Italy	400
Yugoslavia	50
Spain	49
France	6
TOTAL	32,900

resilience and low elasticity as well as warmth of feel and good crease resistance.

Other Saturniid moths have also been used for commercial production of silk, especially the Eri silkworm *Philosamia ricini* (Boisd.), which is bred on castor leaves in India, Indonesia and New Guinea. Wild silk produced by *Antheraea spp.* has long been used in Asia and Africa.

Another insect which has been exploited by man since antiquity is the honeybee. The most primitive communities seek out honeycombs as food. Several species of bees have been cultivated but the most widely found one is the European *Apis mellifera* (L.). The largest honey producer is the United States (100,000 metric tons per annum), followed by Australia (15,000 metric tons) and Canada and the United Kingdom (about 10,000 tons each). The annual production of honey and beeswax in the United States is worth about 48 million US dollars.

While these bee products are useful to man, by far the greatest importance of the bees is as plant pollinators. They play a major role in fertilization of fruit trees, crops and pastures and a decrease in bee population often results in decreased yields. One of the unfortunate consequences of the improper use of insecticides is the deleterious effects these chemicals may have on bees and thus indirectly on plant pollination.

The lac insect *Kerria lacca* (Kerr) produces a hard, tough, amorphous resinous covering which is extracted to form commercial shellac. This scale insect feeds on various species of figs and other trees and is cultivated in India and other Asian countries, especially Thailand in recent years. Shellac has been used in India for 2,000 to 3,000 years but is still a valuable component of varnishes. Although it takes about 15,000 insects to produce one pound of lac, some 90,000,000 pounds of lac are marketed each year at a price of about 50 cents per pound. The annual US consumption is worth about 20 million dollars.

Another commercially important scale insect is the cochineal insect *Dactylopius coccus* Costa. The brilliant red dye produced from the dried and powdered bodies of the insect is used as colouring material in cooking and cosmetics. It was formerly used as a textile dye but in recent years has been replaced by cheaper synthetic dyes. The insect is cultivated in Central America, Spain and Africa.

There are a number of minor uses of insect products and the products of other arthropods, e.g. the use of spider silk for cross-hairs in optical instruments. Some plant galls (which are growths produced by plants in response to the attack of certain insects) are used for the production of dyes and inks. These galls contain large quantities of tannin.

A large number of crops are normally pollinated by bees, flies or moths. These pollinators usually have hairy bodies and move rapidly from flower to flower. The honeybees are particularly well adapted for this and are active in large numbers from early spring to late autumn. Sometimes there is a highly specific relationship between plants and the insects which pollinate them, e.g. the fig wasps which are highly specific pollinating agents for certain species of figs.

The earth would be a very different place without the plant pollinating insects and the insects have played an important part in shaping the evolution of higher plants. It has been suggested that insects have had a major influence on the biochemical evolution of plants and some authors have attributed the presence of essential oils, alkaloids and other repellent substances as an evolutionary means of protecting plants from insect attack.

Another important function of insects in nature is as scavengers, assisting in the destruction of dead animals and plants. Insects initiate the break-up of dead trees and facilitate the entrance of fungi and bacteria which complete the processes of decomposition. This moves reserve materials back into natural mineral cycles. Among the insects boring in dead wood are the termites, and larvae of beetles, flies and Lepidoptera. When some of these insects shift their attention from the jungle to the dead wood used in houses, they become domestic pests. Termites can be pests either through feeding on living wood or on the dead wood used in buildings.

Some insects function in ecosystems as herbivores, others as carnivores, while many are food for other insects or larger animals which in turn are preyed upon by us. Fish, for example, eat mosquito larvae, men eat fish, and adult mosquitoes feed on human blood thus giving a simple food cycle. Arthropods form the primary diet of a large number of vertebrate animals.

They are also eaten by man in some parts of the world. Grasshoppers and locusts have long been important articles of human diet in the Middle East—and they are rich in fats and protein. As well as providing food for man and other vertebrates, insects provide food for other insects of economic importance—especially the predators and parasites of other insects. The sugary excreta ("honeydew") of aphids and scale insects is an important natural food for the adults of many predacious and parasitic insects.

Having considered some of the beneficial points which are often overlooked, we must also consider some of the harmful points. The injuries both to growing plants and stored plant products are enormous and a substantial proportion of all food crops grown are lost to insects each year. In a world which is becoming increasingly short of food as human populations increase, this is a problem which must be viewed more seriously.

Although this book is mainly concerned with insects of agricultural importance in the humid tropics, we cannot overlook the role of insects in the transmission of human and animal disease. Mosquitoes which transmit malaria, filariasis, yellow fever and dengue are responsible for an immense annual toll in human death, suffering and despair. The endemic occurrence of high levels of some of these insect-borne diseases have greatly depressed social and economic development and retarded progress in several developing tropical countries. External and internal parasites and venomous insects also cause direct injury to man and domestic animals. More important are the insects which transmit disease, e.g. the malarial mosquito. Mosquitoes also transmit filariasis, dengue and haemorrhagic fever which are of increasing importance in South East Asia. Many flies carry the microorganisms causing dysentery, skin disease and eye disorders. Insect borne diseases have altered the course of history. Malaria, typhus, yellow fever, sleeping sickness and Chagas' disease have greatly slowed down the economic development of vast areas of tropical America and Africa. There are still many unsolved entomological problems in this field.

FURTHER READING

Anon (1951). *Distribution maps of pests*. Commonwealth Institute of Entomology.
Butler, C. G. (1954). *The world of the honeybee*. London, pp. 226.
Glover, P. M. (1937). *Lac cultivation in India*. Indian Lac Research Inst., Namkum, India, pp. 147.

Metcalf, C. L. and Flint, W. P. and Metcalf, R. L. (1962). *Destructive and useful insects.* 4th edition, McGraw Hill, pp. 1087.

Moreton, B. D. (1969). Beneficial insects and mites. *Min. Agric. Fish. Food Bull.* **20.** 118.

Chapter 2

Classification of Insects and their Arthropod Relatives

The phylum Arthropoda can be divided into four subphyla representing distinct lines of descent.

(a) Subphylum Trilobitomorpha (fossil Trilobites).

(b) Subphylum Chelicerata

      Class Merostomata (horseshoe crabs)
      Class Arachnida (scorpions, spiders, mites)
      Class Pycnogonida (sea spiders)

(c) Subphylum Crustacea

(d) Subphylum Mandibulata

      Class Chilopoda (centipedes)
      Class Diplopoda (millipedes)
      Class Symphyla
      Class Pauropoda

    Superclass Hexapoda

      Class Diplura
      Class Protura
      Class Collembola (springtails)
      Class Insecta.

Since an agricultural or medical entomologist is often consulted about Arthropod pests which are not insects, some of these are considered briefly below.

In the Chelicerata among the class Arachnida, several groups are of economic interest. Of the ten orders of Arachnids, the orders Acarina (mites and ticks), Araneida (spiders) and Scorpionida (scorpions) contain most of the species of economic and medical importance.

In the order Acarina the mites of agricultural importance are found mainly in three families. These are the family Tetranychidae (red spider mites), the family Tarsonemidae (flat mites) and the family Eriophyidae (gall mites).

The red spider mites (which are not necessarily red) are common plant pests and are visible under a hand lens. They feed by puncturing epidermal cells with their stylet mouth parts and the resulting plant damage often appears as a fine pattern of yellow spots on the leaves. On the under surface of the leaf is found fine silken webbing with eggs, nymphs and adults. Tetranychid mites flourish under dry conditions and are serious glasshouse pests in temperate countries as well as field pests in drier parts of the tropics, e.g. *Tetranychus telarius* (L.) which has a wide host range. Hosing or spraying plants with water reduces their attack and acaricidal sprays are often used. Among the most effective natural enemies of red spider mites are ladybird beetles (and larvae) (family: Cocinellidae) and predacious mites of the family Phytoseiidae. In view of widespread resistance to acaricides it is likely that more use will be made of these predators in future.

The Tetranychids have become much more serious pests since World War II following the widespread use of powerful, general insecticides such as DDT which killed off many of the natural enemies but left the red spider mites to reproduce unchecked. Current investigation is seeking more specific acaricides which will kill the pests but not their natural enemies.

The mites of the family Tarsonemidae have broad, flattened bodies and are usually brownish-grey in colour. They are visible under a hand lens. In the adult females the posterior pair of legs are modified into whip-like structures which are readily recognized under the microscope. The hind legs of the males are modified to form rather powerful claspers with a stout terminal claw. Although they are of less economic importance than the Tetranychids, the Tarsonemids are serious pests of some tropical crops and are difficult to control. *Hemitarsonemus latus* (Banks), the broad mite, is a serious pest on tea plants in Papua New Guinea and Ceylon, on *Cinchona* in Sumatra, on *Hevea* in Indonesia, on cotton in Africa and on a number of food and ornamental plants in other parts of the world.

With the exception of the family Eriophyidae the mites have four pairs of legs in the adult stage and three pairs in the first larval stage. The Eriophyids are exceptional in that all stages have only two pairs of legs and these are located at the anterior end of their worm-like, annulate body. The microscopic stylets of the gall mites penetrate single cells of the host. The saliva of certain species has a marked effect on host plant metabolism and results in the production of galls. One species has been incriminated as a cereal virus vector.

Gall mites of the family Eriophyidae are extremely small ($0 \cdot 1 - 0 \cdot 2$ mm in length). A number of species induce the production of abnormal growths (galls, erineum, witches' brooms) on their host plants, while others attack shoots and buds causing extensive damage but no gross plant deformation. Because of their small size these mites are seldom detected without careful microscopic examination of infested plant tissues. In the tropics they are economic pests of citrus, figs, tomatoes, hibiscus and cotton.

Among the largest and most robust of mites are the ticks (suborder: Ixodides) which are all parasitic and occur on a wide variety of vertebrate hosts. They are particularly important as carriers of rickettsial and virus diseases of livestock and of man.

At least three species of small reddish coloured mites of the family Trombiculidae (mokkas or chiggers) are incidental parasites of man and can transmit scrub typhus from rodents to man in South East Asia. Mites of several different families attack man and livestock and cause varying degrees of skin irritation, e.g. members of the family Psoroptidae cause sheep scab, cattle scab and ear infections. The mange mites of the family Sarcoptidae are important skin parasites of man and livestock. The scabies mite, *Sarcoptes scabiei* (DeG.) is a common human skin parasite in the tropics, particularly where hygiene is bad. The immature mites burrow into the skin and cause intense itching, often accompanied by secondary bacterial infection.

One of the most striking characteristics of the Acarina is the lack of apparent body divisions. Abdominal segmentation has disappeared and the abdomen is fused with the prosoma which bears the legs.

In the spiders (order: Araneida) the globular opisthosoma or abdomen is unsegmented but is connected to the prosoma by a very short, narrow waist or pedicel. Silk is produced in spinnerets located posteriorly on the abdomen. The spiders are predators on other arthropods but more work is needed to assess their ecological significance in tropical crops.

Scorpions possess pedipalps, which are large anterior appendages bearing pincers for capturing prey. Their primitive abdomen is composed of a broad, seven segmented preabdomen and a narrow, five segmented postabdomen which ends as a swollen, needle-sharp stinging apparatus containing venom.

The class Crustacea are of interest here mainly because of the order Isopoda which includes the wood lice. These attack plants, especially seedlings, in all parts of the world. The body is dorsoventrally flattened and often bluish-grey or brown in colour. They have seven pairs of legs. The chitinous plates covering the dorsal surface of the body overlap like a coat of armour and some species roll themselves into a ball when disturbed and thus protect their unarmoured ventral surface.

The millipedes of the class Diplopoda are herbivorous, with the exception of a single predacious family. They are seldom serious economic pests but are sometimes of medical interest in that some tropical species can spray a repellant and caustic liquid from a series of repugnatorial glands along the sides of the body. This liquid which can spurt for several centimetres will

Table 2.
Summary of insect classification
(classes or orders in parenthesis are not of economic significance)

SUPERCLASS: HEXAPODA
(CLASS: DIPLURA)
(CLASS: PROTURA)
CLASS: COLLEMBOLA—springtails.
    Order: Collembola

CLASS: INSECTA
SUBCLASS: APTERYGOTA
    (*Order:* Archaeognatha)
    Order: Thysanura—silverfish.

SUBCLASS: PTERYGOTA
    Section: Palaeoptera
    Order: Ephemeroptera—Mayflies.
    Order: Odonata—dragonflies, damselflies.
    Section: Neoptera

    Series 1: Polyneoptera—Blattid/Orthopteroid.
    Order: Blattodea—cockroaches.
    Order: Isoptera—termites.
    Order: Mantodea—mantids.
    (*Order:* Zoraptera)
    (*Order:* Grylloblattodea)
    Order: Dermaptera—earwigs.
    Order: Plecoptera—stoneflies.
    Order: Orthoptera—locusts, grasshoppers.
    Order: Phasmatodea—stick insects.
    (*Order:* Embioptera)

    Series 2: Paraneoptera—Hemipteroid.
    Order: Psocoptera—book lice, bark lice.
    Order: Phthiraptera—lice.
    Order: Hemiptera—bugs, aphids, scales, leaf hoppers.
    Order: Thysanoptera—thrips.

    Series 3: Oligoneoptera—Endopterygota.
    Order: Megaloptera—alderflies.
    Order: Neuroptera—lacewings, ant lions.
    Order: Coleoptera—beetles, weevils.
    (*Order:* Strepsiptera and *Order* Mecoptera)
    Order: Siphonaptera—fleas.
    Order: Diptera—flies.
    Order: Trichoptera—caddisflies.
    Order: Lepidoptera—moths and butterflies.
    Order: Hymenoptera—bees, wasps, ants.

burn the human skin and can cause blindness. The millipedes feed mostly on dead or decaying plant tissue but have sometimes been recorded attacking the underground parts of crop plants. Control measures are seldom required but sprays or dusts of DDT, BHC or diazinon have all been reported effective. Populations may be reduced by removal of decaying plant material from soils.

The centipedes (class: Chilopoda) which have only one pair of legs per segment, compared with the two pairs on each apparent segment on the millipedes, are not plant feeders. They are predacious and carnivorous. They are of medical significance in that they possess poison glands located in the anterior segments of the body and have large poison claws which cover the mouthparts. Many species can inflict a painful bite and some of the larger species are dangerous to man if they are disturbed.

Members of the classes Symphyla and Pauropoda are of no economic significance.

Before considering the insects there are two other animal phyla which sometimes come to the attention of the agricultural entomologist. These are the phylum Aschelminthes, class: Nematoda and the phylum Mollusca. The plant parasitic nematode worms are of great economic importance in both temperate and tropical countries but they will not be considered here.

The phylum Mollusca includes the snails and slugs which are sometimes pests of field crops. Some molluscs are of medical or veterinary significance as intermediate hosts of liver flukes or schistosomes. Of agricultural interest, the giant African land snail *Achatina fulica* Bowdich is a conspicuous and wide-spread pest in several tropical countries. It originated in East Africa but has now spread throughout Asia and the Pacific and can be very destructive to vegetables and young tree crops.

Slugs and snails are generally controlled by using baits containing metaldehyde with bran, sawdust or some other carrier material.

The insect classification used in this book is summarized in Table 2. In general it follows the C.S.I.R.O. (1970). *Insects of Australia*. In the following chapters the insects of major economic importance in in the tropics will be considered in systematic order.

FURTHER READING

Arthur, D. R. (1962). *Ticks and disease*. Pergamon Press, Oxford, pp. 445.
Baker, E. W. and Wharton, G. W. (1952). *An introduction to acarology*. Macmillan, N.Y., pp. 464.
Caswell, G. H. (1962). *Agricultural entomology in the tropics*. Edward Arnold, London, pp. 152.

C.S.I.R.O. (1970). *The insects of Australia*. Melbourne University Press, Melbourne, pp. 1029.

Essig, E. O. (1947). *College entomology*. Macmillan, N.Y.

Grassé, P. (Ed.) (1951). *Traité de Zoologie*, Vols. 6–10. Masson et Cie, Paris.

Imms, A. D., Richards, O. W. and Davies, R. G. (1962). *A general textbook of entomology*, 9th edition. Methuen, London, pp. 886.

Pritchard, A. E. and Baker, E. W. (1955). A revision of the spider mite family, Tetranychidae. *Pacific Coast Ent. Soc. Mem.* 2, San Francisco, pp. 472.

Tuttle, D. M. and Baker, E. W. (1968). *Spider mites of the southwestern United States and a revision of the family Tetranychidae*. University of Arizona Press, Tueson, pp. 143.

Chapter 3

Springtails, Silverfish and some Aquatic Insects

Among the three classes of non-insect Hexapoda the only group of interest here is the class and order Collembola—the springtails, which were formerly regarded as true insects. They are minute, often highly coloured arthropods, very abundant in soil and leaf litter, and very widely distributed. Most species live on decaying plant material and play an important role in the soil biota. A small number of species are agricultural pests and the most serious of these is the lucerne flea, *Sminthurus viridis* (L.). This is particularly a pest of temperate countries where it feeds on grasses, clover and other legumes and causes extensive damage to pastures.

In the Collembola the mouthparts are concealed within the head (entognathous). The abdomen is six segmented and usually bears three pairs of appendages: a ventral tube (collophore) on segment 1, a little hook (retinaculum) on segment 3 and a hooked springing organ on abdominal segment 4. Two suborders can be distinguished, mainly on the basis of their shape and abdominal form. These are:

Suborder: Arthropleona. The body is more or less elongated with the thoracic and abdominal segments distinctly separated, except for the last 2–3 abdominal segments; springing organ absent, e.g. *Onychiurus*.

Suborder: Symphypleona. The body is globular and the thorax and first four abdominal segments are completely fused, e.g. *Sminthurus viridis* (L.), *Bourletiella hortensis* (Fitch).

Although large numbers of Collembola on plants may suggest that they are causing damage, quite often they are secondary invaders feeding on tissues already damaged by other organisms such as nematodes or fungi. Chemical control measures are not normally necessary but organophosphorus insecticides are usually effective if required. In glasshouses their numbers can be reduced by removing decaying plant material.

The Collembola are important components of the soil biota where their role is mainly the reduction of dead plant material. They also feed on fungi, lichens, algae, spores and pollen grains. Some species are abundant in sewage beds where they perform a useful function. Others are found in the intertidal zone. There are relatively few species which are serious agricultural pests.

SUBCLASS APTERYGOTA

Order: Thysanura

In the silverfish the biting mouthparts are externally visible (ectognathous). The flattened, segmented abdomen bears a variable number of lateral and ventral appendages. At the posterior end of the body there are a pair of long cerci and a median segmented process. The order is divided into five families.

Apart from a few members of the family Nicoletiidae these insects do not feed on living plants. They are normally found on dead or decaying plant material such as leaf litter in relatively dry situations. Five species belonging to the family Lepismatidae are domestic pests of cosmopolitan distribution. These are *Lepisma saccharina* L., *Ctenolepisma longicaudata* Esch., *C.lineata* (F.) and *C. urbana* Slab. which live in houses and feed on paper, book-bindings and food scraps. They have a liking for carbohydrates such as paste and glue and are frequently damaging to books. The other cosmopolitan species, *Lepismodes inquilinus* Newm. prefers higher temperatures and is frequently found in kitchens, bakeries or hot rooms. This has led to its common name of "firebrat".

Eggs are laid in secluded places and the newly hatched young resemble their parents. They grow slowly and undergo a large but indefinite number of moults, taking three months to two years to reach maturity, depending upon the temperature and humidity. The body is clothed in shiny scales: hence the name silverfish.

They may be controlled by spraying with the usual household insecticides or by the use of poison baits containing starch or some other carbohydrate as a carrier material and attractant.

SUBCLASS: PTERYGOTA

Section: Palaeoptera

The two orders in this ancient group of insects do not include any major agricultural pests but are of interest for their beneficial role.

Order: Ephemeroptera (Mayflies)

This order can be regarded as entirely beneficial since both the immature forms and the adults are important food for freshwater fish. The adults which

often appear in enormous numbers near lakes or streams are strongly attracted to lights. They may live only a few hours or days as adults (hence the
name Ephemeroptera) but the aquatic nymphs may take one to three years to
reach maturity. The nymphs are almost entirely herbivores or scavengers on
aquatic vegetation and they obtain their oxygen through tracheal gills which
are arranged along each side of the abdomen. The adults have rudimentary
mouthparts and do not feed. The order is somewhat unusual in that there are
two winged stages: the sub-imago which rests for a short period before
moulting to the imago or adult winged form. The adult antennae are short and
bristle-like. The membranous wings are held upwards at rest and the hind
pair much reduced in size. The abdomen ends in a pair of very long cerci,
with or without a median caudal style.

Order: Odonata (Dragonflies and Damselflies)

These insects destroy large numbers of small insects such as flies and mosquitoes as food both in the immature and adult stages. The aquatic nymphs
are excellent food for fishes. The group is thus a beneficial one. On the other
hand there have been several records of the adults feeding on honeybees as they
leave the hive.

The nymphs are found in all kinds of aquatic habitats: both fresh and
brackish and there are a small number of terrestrial species. Odonata have
been found breeding in broken bamboos and Pandanus leaf bases in Malaysia
and in epiphytic Bromeliads in the Americas. Both nymphs and adults are
well adapted to predacious life since the mouthparts are well developed and
the eyes very large and efficient. The nymphs are obligate carnivores and while
they are facultative feeders, exploiting whatever is available, chironomid and
psychodid midges and oligochaete worms appear to be the most favoured
foods. Although both adults and nymphs prey on mosquitoes there does not
appear to be quantitative evidence available to suggest that they are significant biological control agents.

The order is divided into three suborders:

Suborder 1: Zygoptera (damselflies). The fore and hind wings are similar in
shape and venation and their eyes are far apart. The nymphs are slender with
three (or rarely two) large, caudal gills.

Suborder 2: Anisoptera (dragonflies). Here the fore and hind wings are
dissimilar in venation and shape and the rear wings are wider at the base
than the front wings. The eyes touch, or almost touch, each other. The
nymphs are stout and without external caudal gills but possess internal rectal
gills. The antennae have 4 or 6–7 segments. The adults are usually stouter and
more robust insects than the damselflies.

Suborder 3: Anisozygoptera. This small suborder found in Japan and the

Himalayas is intermediate in characters between Zygoptera and Anisoptera. The nymphs have a 5-segmented antenna.

Section: Neoptera

Series: Polyneoptera

Order: Plecoptera (Stoneflies)

These insects are somewhat similar to the mayflies in appearance and also have aquatic nymphs. They differ in having longer antennae and both wings are of equal size or the posterior pairs are larger than the anterior ones. They are of little economic value except as food for freshwater fish, especially trout. As most stonefly nymphs require cool, well aerated water for development they are not commonly encountered at low altitudes in the tropics.

FURTHER READING

Corbet, P. S. (1962). *A biology of dragonflies.* Witherby, London, pp. 247.
Corbet, P. S., Longfield, C. and Moore, N. W. (1960). *A biology of dragonflies.* London, pp. 260.
Needham, J. G., Traver, J. R. and Hsu Yin-Chi (1935). *The biology of mayflies.* Comstock, N.Y., pp. 759.
Paclt, J. (1956). *Biologie den Primär flügellosen Insekten.* Gustav Fischer, Jena, pp. 258.
Salmon, J. G. (1964). An index to the Collembola, Vols. 1 and 2. *Roy. Soc. New Zealand Bull. no. 7*, pp. 644.

Chapter 4   Cockroaches, Mantids and Stick Insects

The cockroaches (order: Blattodea) are familiar domestic pests in all part of the world. They are frequently, but not necessarily, associated with ba hygiene. The order is essentially tropical and is divided into five families, 43 species being known from Australia. They are mostly nocturnal, polyphagou animals and many live in forest litter. Some are wingless while others hav partly or fully developed wings. They undergo an incomplete metamorphosis

The four main cosmopolitan species of domestic importance are:

Blatella germanica (L.)—the German cockroach
Blatella orientalis (L.)—the Oriental cockroach
Periplaneta americana (L.)—the American cockroach
Periplaneta australasiae (F.)—the Australian cockroach.

Some other widespread species are: *Supella supellectilium* (Serv.), *Ischnopter rufescens* (Beauv.), *Rhyparobia maderae* (F.), *Nauphoeta cinerea* (Oliv.) *Neostylopyga* and *rhombifolia* (Stol.).

The cockroaches feed on human food but they are not necessarily of majoi importance as vectors of disease (except in hospitals where they may hav access to infectious material). They are difficult to eradicate by insecticides s application must be thorough and persistent. Hygiene is important.

The eggs are laid in typical egg cases (oothecae). In *B. germanica* the egg cases are carried around by the female for some time before they are de posited. On the other hand *B. orientalis* and *P. americana* drop their egg cases as soon as they are produced. The cockroaches are relatively slow-growing and long-lived insects but there is considerable variation between species Thus at 25° *B. germanica* takes about four months to reach maturity while *B. orientalis* and *P. americana* spend 18–20 months in the immature stages, and the adults live a further 5–15 months at this temperature.

The mantids (order: Mantodea) are beneficial insects since they are carni vorous and destroy many harmful insects. It is essentially a tropical group. The forelegs are strongly developed for catching and holding insects and other small animals. The praying mantis often stands still for long periods with the forelegs held out in front of the head as if in prayer (hence the popular name "praying mantis"). The eggs are laid in oothecae of character istic appearance and these are cemented onto twigs, branches, walls, etc.

lantid populations tend to be small because of egg parasitism and predation ⚏n the early stages so the group is only of minor economic importance. *Iantis religiosa* L. has been introduced into Canada as a predator of pasture ⚏sects and it has been reported to feed extensively on grasshoppers, crickets ⚏nd beetle larvae.

The stick and leaf insects of the order Phasmatodea include some of the ⚏rgest insects known, some being over a foot in length. The leaf insects of ⚏e family Phylliidae have broad, flat bodies with plate-like expansions on the ⚏gs. Their resemblance to leaves in colour, shape and "venation" is remark- ⚏ble. The stick insects of the family Phasmatidae have very long, slender ⚏odies, sometimes spiny and with lateral expansions on the legs or abdomen. ⚏hey closely resemble twigs or branches and when disturbed frequently feign ⚏eath. Some phasmids possess chromatophores which enable them to change ⚏olour to match their background. Phase differences have also been reported ⚏n some species which reach high population densities. Among these, *Pod-
⚏canthus wilkinsoni* Macl. and *Didymuria violescens* (Leach) are sometimes ⚏erious pests of Eucalyptus forests in Australia where they cause extensive ⚏efoliation. The eggs, which are often beautifully ornamented, are laid singly ⚏nd fall to the ground where they may remain dormant for some time. The ⚏ymphs and adults are all phytophagous. This order is particularly well ⚏eveloped in South East Asia and in Central and South America.

FURTHER READING

Cornwall, P. B. (1965). *The cockroach*, Vol. 1. William Heinemann, London, pp. 391.

Chapter 5   *Termites*

Order: Isoptera

These social insects live mainly in tropical and subtropical regions in large colonies and have several morphological forms within each species (polymorphic). The occurrence of these densely populated societies has given rise to the popular name of "white ants" but this is unfortunate as they are only remotely related to the ants which belong to the order Hymenoptera. The immature ant larvae are inactive, legless forms which are tended by the workers but the immature nymphs of Isoptera are active and of the same general form as the adults. They attain maturity after about seven moults.

Each colony contains several different castes (morphs) which differ in appearance, structure and function. They can be divided into two groups (a) *reproductive castes* and (b) *sterile castes*. The reproductive castes comprise the winged sexual forms (primary reproductives) which establish new colonies after a mating flight, the sexual pair being the king and queen of the new colony. The wings are shed after the mating flight. Should the king and queen die their place is taken by replacement reproductives which do not have properly developed wings. The two most important of the sterile castes are the soldiers and workers which develop from eggs laid by the primary reproductives. Although these are terminal stages of development they can be regarded as permanently juvenile forms (of either sex) and incapable of reproduction. The workers are soft, whitish, wingless and blind. Their functions are to build and clean the termitaria and to look after the young and the reproductive forms, and to forage for food.

The soldiers are usually more specialized in structure: often with large heads and powerful projecting mandibles. Their function is to defend the colony, especially from ants. In the family Termitidae some of the soldiers have the head modified into a long pear-like structure. At the end of the head of these nasute forms there is a duct for the discharge of a sticky, repellent fluid which is used both for defence and for nest construction.

Many types of nests are encountered amongst the termites. Some termitaria are large and complex structures above the ground. Some species have complex underground nests with special gardens for growing fungus food. Others nest on or in trees or have simple underground nests. In the more primitive wood-feeding species the worker caste is lacking and the nest is simply a series of galleries excavated in the wood. Some ground-dwelling species enter wood through the soil. These may be very injurious to the woodwork of buildings in contact with the ground. Sometimes the termites come above the surface of the ground to attack buildings nearby. They protect themselves with a covered passageway of mixed earth and faecal material since the workers are usually quite intolerant of light or air movement. This arrangement also provides protection from enemies and maintains a suitable humidity. Sometimes long runways extend from the ground to the tops of buildings or high trees. Some of the groups which nest above ground feed extensively on the roots of grasses and crops.

However, not all the activities of termites can be regarded as harmful. Some species increase the aeration and drainage of tropical soils and may play an important role in soil building. Termite mounds are rich in mineral salts and organic matter. In parts of Africa *Macrotermes* plays a role in the conversion of savanna to forest. Termites are able to digest cellulose with the assistance of symbiotic protozoa and bacteria and thus play an important role in recycling plant nutrients. They are also the basic food for many migratory insectivorous birds, as well as ant eaters, pangolins and ant bears. They are eaten by man and in parts of Africa their oil is used for cooking.

The Isoptera are divided into six families of which three are of major economic significance.

Family 1: Mastotermitidae

This includes a single species *Mastotermes darwiniensis* Frogg. which is confined to tropical Australia and New Guinea. This species normally nests in trees and stumps but can be very destructive to buildings as well as trees and crop plants including sugar cane.

Family 2: Termopsidae

These damp wood termites are usually found in standing trees or fallen logs. Very few are pests.

Family 3: Hodotermitidae

These harvester termites are found in drier parts of Africa and the Middle East. They forage in columns above ground for grasses which are cut and stored in underground nests. They are of little economic significance.

Family 4: Kalotermitidae

The dry wood termites include a number of species causing severe damage in buildings and furniture, e.g. *Cryptotermes domesticus* (Hav.) which is widespread in the South Pacific and *C. brevis* (Wlk.) which occurs in Central and South America, the West Indies, Africa and the Eastern Pacific. The form galleries inside the wood and make occasional holes through which excreta are ejected.

Family 5: Rhinotermitidae

These moist wood termites include a number of species causing severe damage to timber and to living trees. In New Guinea *Coptotermes elisae* (Desn.) attack hoop pine (*Araucaria cunninghamii*). The huge nests are mainly below ground and associated with tree roots. External runways covered with mud can be seen on the trees but there is also extensive tunnelling in the wood which eventually kills the trees. Within plantations several trees may be attacked simultaneously from the same nest. Control is difficult and involves digging up the nest and destroying the queen. Other species of *Coptotermes* are the most damaging group of termites in Australia. Some are mound builders while others are not.

Family 6: Termitidae

This is the largest family of termites. They are mainly wood or grass eaters with subterranean habits and many are mound builders. There are six sub families. *Microcerotermes biroi* (Desn.) frequently builds nests in the coconut trees in Papua New Guinea, the Solomon Islands and Samoa. It also attacks cocoa and rubber (Figs 1 and 2). *Macrotermes bellicosus* (Smeath.) is a pest of ground nuts in tropical africa. It destroys the collar of the plant at ground level. This species builds conspicuous mound nests in the African savanna *Microtermes thoracalis* Sjost., the cotton soil termite, feeds in the central tissues of the collar of the tap-root of various crops in Africa (including cotton, groundnuts, cereals and tomatoes). This species lives in small subterranean nests. The date palm termite, *Odontotermes smeathmani* (Full.), a poly-phagous species, forms galleries over the young leaves of date palms and also bores into the base of the stems of older trees.

The control of termites requires a different approach according to the identity of the termite and the nature of their food. It is obviously more economical to prevent termite attack than to control existing infestations which may be concealed within walls or tree trunks. There are three basic methods of preventing termite attack on buildings: wood preservation, site poisoning and constructional measures and it is necessary to determine which method or methods are most appropriate. Wood preservation is usually carried out on the green timber or during the curing process. This is probably

Fig. 1. Termite damage to cocoa trunk.

Fig. 2. Runways of termites (family: Termitidae)
on cocoa trunk.

the most economical method of control in the tropics—providing the chemical treatment used is effective against the local termites.

Constructional measures aim at denying soil dwelling termites access to buildings. These methods are ineffective against dry wood termies which do not require contact with the ground. However, even where dry wood termites are the dominant pests it is usually worthwhile incorporating barriers in new buildings as the ground dwellers are swift and damaging and often occur as well as dry wood species in most areas.

Site treatment involves the use of a barrier of insecticide-treated soil around the foundations of a building. Some of the organochlorine insecticides have remained effective for 10–14 years after one soil treatment. Soil treatment is ineffective against dry wood termites which often gain access to buildings in infested furniture. The use of a fluoridated silica aerogel dust in wall and ceiling cavities has proved to be a promising control which can be applied economically.

Termite control is particularly important in forestry when Eucalyptus is being established in semi-arid areas. It has been found that the incorporation of 2% dieldrin dust in potting soil at a rate of 10 ounces per cubic yard gave long term protection to the seedlings. Control of infestation in crops has usually been through soil application of insecticides, and through the elimination of nests where this is feasible.

In recent years a new problem has arisen with termite damage to plastics, especially PVC, polyethylene and cellulose esters. The use of non-toxic mineral dusts such as zircon flour and hard silicon dust added to the plastic filler have been effective and more desirable than the use of insecticides which, though effective, could have introduced possible toxic hazards.

The treatment of infested buildings usually involves surface treatment, fumigation and dusts. Surface treatment involves painting insecticidal emulsions (such as dieldrin + pentachlophenol) or application of oil based chlorinated hydrocarbon sprays onto exposed wood surfaces. Fumigation, using methyl bromide for example, is effective but requires a gas-tight building. The use of dusts following fumigation is useful to prevent reinfestation by drywood termites.

FURTHER READING

Grassé, P. P. (1949). *Traité de Zoologie*, Vol. 9, *Insecta*. Masson et Cie, Paris.
Gray, B. (1972). Economical tropical forest entomology. *Ann. Rev. Ent.* **17**: 313–354.
Harris, W. V. (1961). *Termites, their recognition and control*. Longmans, London.
Harris, W. V. (1965). Recent developments in termite control. *PANS* (A)**11**: 33–43.
Harris, W. V. and Sands, W. A. (1965). The social organization of termite colonies. *Symp. Zool. Soc. Lond. no.* 14, pp. 113–131.
Howse, P. E. (1970). *Termites: a study in social behaviour*. London, pp. 150.
Krishna, K. and Weesner, F. M. (Eds) (1969). *Biology of termites*, Vols 1 and 2, Academic Press, London, pp. 598 and 643.
UNESCO (1962). Termites in the humid tropics. *Proc. New Delhi Sympos.* 4–12 *Oct.* 1960, pp. 259.

Chapter 6   *Orthoptera, Dermaptera and Some Paraneoptera*

The grasshoppers and related insects are of great economic importance, especially in arid regions. They are plant eaters with well developed chewing mouthparts. Mainly medium to large sized insects, they often have powerful hind legs with femora enlarged for leaping. They may be solitary or gregarious. The order is divided into two suborders and six superfamilies, three of which are of economic interest.

Order: Orthoptera

Suborder: Ensifera

Here the antennae have more than 30 segments and the auditory organs (where present) are located on the fore tibiae.

Superfamily 1: Gryllacridoidea

Although this superfamily includes some interesting giant cave crickets and wetas, they are not of economic interest.

Superfamily 2: Tettigonioidea

This includes the long-horned grasshoppers and katydids. The antennae are usually longer than the body. They feed on the leaves and stems of herbaceous and woody plants. Among the 5,000 species of this superfamily are a number of minor horticultural and agricultural pests.

Superfamily 3: Grylloidea

These are the true crickets and include both the mole crickets and the tree crickets. The tarsi are three-segmented. Apart from the tree crickets they live under logs or stones during the day and emerge to feed and stridulate at night. The black field crickets (family: Gryllidae) do some damage to pastures and crops. The song of crickets is a familiar night sound in the tropics and in parts of Asia crickets are kept as pets (for singing or fighting). The mole crickets (family: Gryllotalpideae) have the front limbs greatly modified for digging with flattened femur and tibia. They make permanent burrows with galleries. Although only about 50 species are known they are widely distributed through the tropics. *Gryllotalpa africana* Beav. is a minor crop pest which stores seeds underground.

Suborder: Caelifera

The antennae have less than 30 segments and the auditory organs (where present) are located on the first abdominal segment.

Superfamily: Acridoidea

In the short-horned grasshoppers the antennae are shorter than the head and thorax, there are three tarsal segments and the ovipositor is short and inconspicuous. The wings, which are often brilliantly coloured, are folded fan-wise beneath the tegmina when at rest. Stridulation is produced by rubbing a scraper on the tegmina against a file on the femur or by rubbing the hind wings against the thickened veins of the tegmina.

Among the 10,000 species of this superfamily are the most destructive Orthoptera including the locusts whose migratory swarms have brought desolation to many parts of the tropics from time to time. The desert locust, *Schistocerca gregaria* Forsk. is a long distance flier around the Mediterranean and its ravages range from North Africa to Northern India. It sometimes occurs in enormous populations, e.g. in Cyprus in 1861 it is reported that 1,300 tons of eggs were destroyed.

The migratory locust, *Locusta migratoria* (L.), occurs in Europe, Asia, Africa, Madagascar, New Guinea, the Pacific, Australia and New Zealand. It breeds in dry grassy places near jungles and swamps and also in burnt off areas where the vegetation is sparse (Fig. 3).

The red locust, *Nomadacris septemfasciata* (Serville) occurs in South and South East Africa. A recent outbreak which commenced in 1930, spread over almost the whole of Africa south of the Sahara. The principal plague locust in dry parts of Australia is *Chortoicetes terminifera* Walk.

For centuries the locusts have been known as great agricultural scourges. Since they feed mainly on cereals and grasses and appear in enormous populations, their attacks may result in famine for man and domestic animals. A great deal of scientific study has been devoted to locusts. It has been found that the plague species exist in two forms or phases which differ in colour, appearance and behaviour. In the *solitary phase* the sparse populations behave much like other grasshoppers. However in the *gregarious phase* aggregation occurs and dense swarms build up and migrate. The behavioural stimuli resulting from crowding of the nymphs brings about the physiological changes leading to the gregarious phase. Under appropriate weather conditions, the build up of nymphs occurs in *outbreak areas* of characteristic topography, soil and vegetation characteristics for each locust species.

For many years the Antilocust Research Centre has studied locust outbreaks. An important aspect of locust control is to detect the appearance of gregarious phases in known outbreak areas—particularly in Africa and the Middle East where they are well defined for different species.

Fig. 3. The migratory locust: *Locusta migratoria* (L.).

In the case of the migratory locust *L. migratoria*, an important African outbreak area is a bend in the River Niger near Timbuktu. Outbreaks occur only in some years, thus one occurred in 1928 and had collapsed by 1933. With the desert locust, *S. gregaria*, the migration of swarms is an annual event and there are no well defined outbreak areas. The outbreak areas for the Australian locust, *Chortoicetes*, lie in remote country along the border between the states of New South Wales and Queensland and migrating swarms often move across state borders.

The South American locust, *Schistocerca americana* (Drury) formerly caused extensive damage in Central and South America where it was a permanent pest. This species has largely been brought under control by the work of the Food and Agricultural Organization of the United Nations working from Managua in Nicaragua.

Locusts have long been a popular human food, particularly in Middle Eastern countries. They are rich in oils and proteins and in many places form a regular part of the diet.

Control is generally effected by poisoning with baits or sprays, especially in the nymphal stage and before the change to the gregarious phase has taken place. Locusts take poison baits readily and dying and dead locusts are often eaten by their fellows. BHC is widely used in bran baits. Although it is much more efficient to destroy nymphs at their breeding sites before swarms have formed, this demands an extensive organization able to carry out regular field surveys for evidence of mass breeding—often in remote areas. The antilocust organizations of the Middle East and Central America have performed a great service in this direction.

Order: Dermaptera

These insects are readily recognized by the modified cerci which form heavily sclerotized forceps at the hind end of the body. They are nocturnal animals and mostly plant feeders though some tropical species are carnivorous or cannibalistic. Two species occurring in Hawaii are important in the biological control of cicadellid pests.

Earwigs are seldom of economic importance unless they occur in large numbers, when they may be occasional pests of fruit or vegetation. The European *Forficula auricularia* (L.) is a plant pest. Dermaptera should be preserved in alcohol as some internal characters are frequently used for species determination.

Series: Paraneoptera (Hemipteroid insects)

Order: Psocoptera (Book Lice, Psocids, Corrodentia)

Many species live on bark and plant debris while others are minor pests in libraries and laboratories where they eat paper, glue, books and dried insects. They feed mainly on moulds and spores. Control is facilitated by reducing the humidity which will lessen the growth of fungi and make it difficult for the psocids to find food. They are sometimes found in very large numbers in warehouses and granaries, particularly in mouldy cereal products.

Order: Phthiraptera (Lice)

These external parasites of birds and mammals spend their entire life on the host. The order is divided into three suborders: Mallophaga, Anoplura and Rhynchophthirina. The Rhynchophthirina includes a single species occurring on elephants and will not be considered further.

The suborder Mallophaga (biting lice, bird lice) are found on birds and some mammals and include several species of agricultural importance. The head of the adult is relatively large and possesses biting (mandibulate) mouthparts. The eggs are laid separately and attached to the bases of feathers or hairs of the host by a gluey secretion. The nymphs which resemble the adults in appearance, hatch in about four days and mature in 1–4 weeks while the adults live for a few months.

Many of these insects are highly host specific and each host may be parasitized by 5–6 species of lice. They feed on feathers, hair, skin, scales and sometimes around wounds. When claws are present they are frequently modified for clinging to hair or feathers. Another adaptation to parasitic life is the wingless condition of the adults.

About 2,000 species are known and while they are not blood-suckers or vectors of disease, in large populations they may cause considerable discomfort to their host. The common fowl louse *Menacanthus strumineus* (Nitsch.) bites through quills to obtain blood and also bites through the skin of the host. *Menopon gallinae* (L.) is another widely distributed pest of fowls. Other economic species include *Damalinia spp.* attacking cattle, sheep, goats and horses, *Trichodectes canis* (DeG.) on dogs and *Felicola subrostratus* (Burm.) on cats. They are usually controlled by DDT or BHC dusts and are not serious pests though egg production of fowls may be depressed and young chickens are occasionally killed by chicken head lice.

The dust baths taken by birds afford a natural means of louse control.

The suborder Anoplura (sucking lice) are parasites of mammals and can be distinguished from Mallophaga by their relatively smaller head, the presence of piercing and sucking mouthparts and the location of the thoracic spiracles. These are dorsal in Anoplura and ventral in Mallophaga. The mouthparts are adapted for piercing the skin and sucking blood.

In the family Linognathidae members of the genus *Linognathus* parasitize dogs, sheep, goats and cattle. In the family Haematopinidae are a number o, pests of ungulates and *Haematopinus spp.* are found on cattle, pigs, horses and buffalo. Many of the family Hoplopleuridae are parasites of rodents including *Polyplax* spp. on rats and mice. *Polyplax spinulosa* (Burm.) can transmit murine typhus among rats but is a less important vector than rat fleas.

The three human lice: the head louse, the body louse and the crab or pubic louse are members of the family Pediculidae which is parasitic on Primates. These species are:

Pediculus humanus capitis DeG.—the head louse
Pediculus humanus humanus L.—the body louse
Phthirus pubis (L.)—the pubic or crab louse.

The two genera are readily distinguished since *Phthirus* is shorter, rounder and crab-like but the subspecies of *Pediculus humanus* are difficult to separate morphologically.

The head louse is normally found on the scalp where the eggs are cemented to individual hairs in a characteristic manner. These eggs, whether hatched or not, are termed nits. The nymphs which resemble the adults in appearance pass through three instars before reaching sexual maturity. Adult lice are dirty-white to greyish brown in colour and the terminal claws are modified for grasping hairs.

The body louse has a similar life history but eggs are glued to the fibres of the clothing rather than to hairs. Body lice spend most of their time in the clothing, particularly in undergarments, and tend to congregate along the seams. They visit the skin to feed when the host is sitting or lying quietly.

The crab louse is usually confined to the pubic region but occasionally in heavy infestations it is found in the armpits or even in the eyebrows. These lice tend to remain in one place, clasping the pubic hairs with their legs. This species is unimportant as a vector of disease.

On the other hand *Pediculus* is a vector of relapsing fever, trench fever and epidemic typhus and indeed it has altered the course of history through this role. Under crowded and insanitary conditions epidemics may result as the lice move and transmit disease organisms from one person to another. Major epidemics have particularly occurred in wartime when standards of hygiene are low and large numbers of troops are brought together under unsatisfactory conditions.

In mediaeval times lice were regarded as a mark of holiness but today they are regarded as distasteful evidence of bad hygiene. They are all too common, both in developing countries and in poorer urban areas of developed countries. They may be controlled by the application of insecticides to the hair or, in the case of the body louse, by application of insecticides to the clothing. Clothing may be disinfested by heat sterilization or chemical fumigants. Personal and domestic hygiene are obviously important.

Order: Thysanoptera (Thrips)

These tiny insects are readily recognized by their delicate, narrow wings with greatly reduced venation and long marginal setae. The asymmetrical, biting, rasping mouthparts are used for superficial feeding on soft, recently formed plant tissues: mainly leaves, flowers and fruit. The damage caused by thrips is very characteristic and often manifests itself as superficial scars accompanied by numerous brown faecal spots on the tissue which is being attacked.

There are two suborders: the Terebrantia which have at least one complete longitudinal vein in the wings and a saw-like ovipositor in the females; and the suborder Tubulifera where the wings are almost devoid of venation, the females lack an ovipositor and the terminal segments of the abdomen form an extensible tube for depositing eggs.

Although there is a rich thrips fauna in the Australian and Oriental regions, relatively few species are of economic importance. The cosmopolitan *Thrips tabaci* L., which transmits the spotted wilt virus of tomatoes, is also a serious pest of onions and attacks a wide range of host plants. The polyphagous *Heliothrips haemorrhoidalis* (Bch.) is also a cosmopolitan pest and is frequently encountered in greenhouses. The red banded cocoa thrips, *Selenothrips rubrocinctus* (Giard) is found wherever cocoa is grown: feeding on leaves, flowers and developing fruit. Severe attack may result in defoliation or damage to young developing pods. It is worst in dry conditions and in unshaded plantations. This species also attacks cashew, mango and other hosts. Other species of thrips attack coffee, cotton, citrus and cereals.

FURTHER READING

Anon. *The locust handbook*. Antilocust research centre, London. (Continuing publication).

Baer, J. G. (1957). Premier symposium sur la spécificité parasitaire des parasites de Vertèbres. Neuchatel, pp. 324.

Ferris, G. F. (1951). *The sucking lice*. San Francisco, pp. 320. (Mem. Pacific Coast Ent. Soc. **1.**)

Pessen, P. (1951). Ordre des Thysanopteres *in* Grassé, P. (Ed.) *Traité de Zoologie* Vol. 10. Masson et Cie, Paris, pp. 1805–1809.

Rothschild, M. and Clay, T. (1952). *Fleas, flukes and cuckoos*. Collins, London, pp. 304.

Uvarov, B. P. (1928). *Grasshoppers and locusts*. Cambridge University Press, London, p. 481.

Chapter 7   *Hemiptera*

This old and important order is divided into two suborders: the suborder Homoptera where both wings are of uniform consistency, and the suborder Heteroptera where the forewings are basally thickened to form hemelytra with a membranous apex. In both suborders the mouthparts are adapted for piercing and sucking. The mandibles and maxillae form slender stylets containing a suction canal and a salivary canal. The stylets lie within the fleshy, dorsally grooved labium. Metamorphosis is gradual (Exopterygota) and the nymphs resemble the adults.

This large and successful order contains many members of agricultural importance and a few of medical importance.

Suborder: Homoptera

The forewings are of uniform texture and are usually held roof-like over the abdomen. The labium is inserted close to the prosternum. They are never aquatic. The suborder is divided into nine superfamilies.

Superfamily 1: Peloridoidea

This is a small group of temperate moss and liverwort dwellers of no economic significance.

Superfamily 2: Fulguroidea

These plant hoppers mostly feed on the vascular tissues of plants. The nymphs of several groups produce masses of white, flucculent wax from the hind end of the body and many hop when disturbed.

Family: Delphacidae

This family includes a number of vectors of virus diseases of graminaceous plants, e.g. *Perkinsiella saccharicida* Kirk., a native of North Queensland which has spread through South East Asia, Hawaii, South America and Africa. This is a vector of Fiji disease virus of sugar cane. The eggs of *Perkinsiella* are inserted into the cane stems and are easily spread accidentally by man during transplanting. An egg parasite gives good control in Australia. The main Delphacid virus vectors are listed in Table 3. As well as transmitting

Table 3. Some Delphacid vectors of plant virus diseases

Species	Host	Virus	Distribution
Laodelpha striatella (Fall.)	Maize	Maize rough dwarf virus	Europe, Israel
Javesella pellucida (Fab.)	Maize	Maize rough dwarf virus	Europe
Peregrinus maidis Ashm.	Maize	Maize (corn) mosaic virus	West Indies, Hawaii, Tanganyika
Perkinsiella saccharicida Kirk.	Sugar cane	Fiji disease virus	South East Asia, Pacific, South America, Africa
Perkinsiella vastatrix Breddin	Sugar cane	Fiji disease virus	Philippines
Riboautodelphax albifasciata (Mats.)	Rice	Black streaked dwarf and stripe disease	Japan
Unkanodes sapporanus Mats.	Rice	Black streaked dwarf and stripe disease	Japan
Nilaparvata lugens (Stål)	Rice	Grassy stunt	India, Ceylon, South East Asia, New Guinea, Pacific
Sogatodes cubanus (Crawf.)	Rice	Hoja blanca	Central and South America
Sogatodes orizicola (Muir)	Rice	Hoja blanca	Central and South America

virus diseases some Delphacids cause direct damage to their host plants, e.g. *Sogatella furcifera* (Horvath) which attacks rice and various grasses in India, South East Asia, Japan and the Pacific and *Saccharosyne saccharivora* (Westm.) which attacks sugar cane in Central and South America and the West Indies. The taro leaf hopper (*Tarophagus proserpina* (Kirk.)) occurs through South East Asia, Australia and the Pacific, on *Colocasia* spp and *Alocasia* spp.

Family: Ricaniidae

These medium sized insects with triangular wings are often brown and white in appearance. The passion vine leaf hopper, *Scolypopa australis* Walk. causes extensive damage to passion vines as well as ornamental and food plants in Australasia and New Guinea. Its copious honeydew provides food for other insects as well as a medium for the development of sooty mould fungi. Cases of poisoning have been reported from honey collected by bees from the honeydew of *Scolypopa* feeding on the poisonous plant *Coriaria* in New Zealand.

Superfamily 3: Cercopoidea

Family: Cercopidae (Spittle Insects, Cuckoo Spit, Froghoppers)

The larvae of several genera surround themselves with a frothy mass resembling saliva but derived from excreted material mixed with air bubbles. This reduces their water loss and conceals them from their enemies. Though they are well represented in the tropics and are all plant feeders they are of little economic significance apart from the sugar cane froghopper of the West Indies and South America: *Aeneolamia varia saccharina* Dist. This was for many years the most serious pest of sugar cane in Trinidad and caused more injury than all the other pests together. Eggs are laid in dead cane leaves or even in the soil. The nymphs feed only on sap from the roots which they reach through cracks in the soil. Masses of froth form over colonies of feeding nymphs. The black and yellow winged adults suck sap from the canes but do not form froth. Extensive feeding activity results in withering of the leaves and stunting of the stalk—the so called "froghopper blight", which appears to be caused by toxic adult saliva rather than by loss of nutrients through feeding. The severity of the damage is influenced by weather and soil conditions.

Superfamily 4: Cicadoidea

The cicadas (family: Cicadidae) are abundant in the tropics and are conspicuous because of their well developed sound-producing organs. The adults

Table 4. Some Cicadellid vectors of plant virus diseases

Species	Host	Disease	Distribution
Nephotettix nigropictus (Stål) (syn. *N. apicalis* (Motsch.)	Rice	Dwarf virus, yellow dwarf mycoplasma, transitory yellows virus, yellow-orange leaf virus	India, Ceylon, South East Asia, New Guinea, Caroline Islands
Nephotettix cincticeps (Uhl.)	Rice	Dwarf virus, yellow dwarf mycoplasma, transitory yellows virus	Japan
Nephotettix virescens (Dist.) (syn. *N. impicticeps* Ish.)	Rice	Yellow dwarf mycoplasma, tungro, leaf yellowing, Penyakit merah, yellow-orange leaf	Japan, India, South East Asia
Recilia dorsalis (Motsch.)	Rice	Dwarf, orange leaf	Japan, Philippines
Draeculacephala portola Ball.	Maize	Chlorotic streak	Americas, Java, Queensland, Taiwan
Cicadulina mbila (Naude)	Sugar cane and maize	Streak disease virus	Africa
Cicadulina zeae China	Maize	Streak disease virus	Africa
Cicadulina nicholsi	Maize	Streak disease virus	Africa
Cicadula bimaculata (Evans)	Maize	Wallaby ear virus	Queensland

insert their eggs into the bark of trees, sometimes causing minor damage to fruit trees. The nymphs hatch and move into the soil where they feed on roots, sometimes for very long periods before reaching maturity. They are, however, of little economic significance.

Superfamily 5: Cicadelloidea

Family: Cicadellidae (Jassids, Leaf hoppers)

This large family of rather small insects includes a number of important vectors of plant virus and mycoplasma diseases. Some species also cause severe crop damage through their feeding activities and injection of toxic saliva (e.g. *Empoasca* spp. on cotton in Africa). *Empoasca* spp. also attack lucerne in Queensland, potatoes in America, and tea in Ceylon. Some major virus vectors on rice and maize are listed in Table 4. Some of the Cicadellids have a limited host range but others, such as *Macrosteles fascifrons* (Stål.), the vector of aster yellows mycoplasma, have a wide range of host plants. Groundnut mosaic virus in Java is transmitted by *Orosius argentatus* Evans.

Members of the family Membracidae (tree hoppers) have a massive and bizarre enlargement of the pronotum. They are of little economic significance.

Superfamily 6: Psylloidea

Family: Psyllidae (Chermidae, Jumping Plant Lice, Lerp Insects)

These tiny insects resemble miniature cicadas and also somewhat resemble aphids but have a different wing venation. The nymphs are very characteristic being flattened, almost scale-like insects with large wing pads and often a marginal fringe around the body. Some gall forming nymphs are completely embedded in host plant tissue while others form shallow pit galls on leaves. Some species secrete an ornamental cover (lerp) which conceals the developing nymph. Some of the pest species have toxic saliva, e.g. *Psylla* spp. on apples and pears. Psyllids are pests of cotton, cacao, citrus and timber trees, e.g. *Trioza erytreae* (DeG.) the citrus psyllid of Africa. *Diaphorina citri* Knn. transmits the virus causing citrus leaf mottle yellows disease in the Philippines.

Superfamily 7: Aphidoidea

The role of aphids as the main vectors of plant virus diseases makes this group the most important superfamily in the Hemiptera from the economic point of view. There are some 170 plant viruses transmitted by 102 species of aphids but some aphid species can transmit a large number of different viruses—e.g. *Myzus persicae* (Sulz.) which can transmit 108 different plant viruses. Many aphids feed on the vascular tissues of plants and this is frequently an important factor in the transmission of disease. Most of the aphid-transmitted viruses are of the non-persistent type which can be aquired during a short feed and then transmitted at once but the insect remains infective for only a few hours or days, and infectivity declines with each successive feeding or probing activity. This would imply that the role of the aphid is largely one of mechanical transmission, however the substantial differences between aphid species in their efficiency of transmission of the same virus suggest that the situation is more complex than simple mechanical transmission. A few viruses can be transmitted by aphids for a longer period and in a very few cases there is evidence of virus multiplication within the aphid (persistent viruses). However, the majority of persistent viruses and mycoplasmas are transmitted by leaf hoppers.

The aphids are highly polymorphic and within one species there may be winged and wingless forms and sexual and parthenogenetic forms. This polymorphism satisfies the conflicting needs for rapid exploitation of their hosts (perhaps during a brief growing or flowering period), efficient dispersal and host finding, and survival during unfavourable periods.

In the tropics aphid reproduction is usually continuously parthenogenetic since the low temperatures combined with short day lengths required for the production of sexual forms in many species do not occur close to the equator. Aphids have a high reproductive rate. An adult can produce about 80 young, usually born alive at a rate of about 2–5 per day and these become reproductive in 10–15 days. This rapid parthenogenetic reproduction allows efficient exploitation of the host plant during optimum plant growth conditions.

When populations become dense and the quality and quantity of food declines, winged forms are produced and leave the plant after their last moult. These alates are the main dispersal forms and also the main vectors of plant viruses, which may be carried from the original host plant or (more usually) picked up on route during a series of short flights before the alate settles down to reproduce.

Some species cause plant damage through their toxic saliva but most plant damage is caused directly through the feeding activities of enormous aphid populations which can develop under favourable conditions. Aphid honeydew, which is a watery mixture of excreta and undigested plant sap, is rich in sugars, amino acids and mineral elements and is an important food for parasitic (and other) Hymenoptera.

Among the aphid species of particular economic significance in the tropics are:

Aphis gossypii Glover, the cotton aphid which is also a pest of cucurbits.

It transmits cucumber mosaic virus which affects a wide range of host plants. It also transmits virus diseases of sugar cane, cotton, groundnuts, papaya, sweet potatoes and citrus.

Brevicoryne brassicae (L.) the cabbage aphid attacks cruciferous plants and carries their viruses in most parts of the world.

Myzus persicae (Sulz.) has a wide range of host plants (polyphagous) and transmits many virus diseases of tropical crops.

Pentalonia nigronervosa Coq., the banana aphid, transmits banana bunchy-top virus of bananas and manila hemp.

Rhopalosiphum maidis (Fitch) is confined to plants of the grass family and transmits sugar cane mosaic and maize leaf fleck virus.

Sipha flava (Forbes) is a pest of sugar cane in the West Indies and Central America. It causes direct damage through its feeding activities but does not appear to be a virus vector.

Toxoptera citricidus (Kirk.) which occurs on citrus and other Rutaceae transmits the important citrus tristeza virus and several other citrus viruses.

Toxoptera aurantii (B. de F.) occurs on citrus, cocoa, coffee, tea, cola and other hosts. It also transmits citrus tristeza disease.

Superfamily 8: Aleyrodoidea

Family: Aleyrodidae (Whiteflies)

The immature stages of these tiny insects resemble coccids because they do not move about on the leaf after their first instar. The winged adults have a white waxy powder covering their body.

They are important glasshouse pests but also cause feeding damage to a variety of tropical crops including citrus, coconuts, cotton, sugar cane, tobacco and cassava. *Dialeurodes citri* (Ash.) is the citrus whitefly of India, Ceylon, Vietnam, China, Japan and the Americas. It also attacks coffee, Gardenia and other ornamental trees and shrubs. The orange spiny whitefly, *Aleurocanthus spiniferus* Quaint. attacks citrus, *Rosa* spp. and fruit trees in South East Asia, India, Ceylon and East Africa. The citrus blackfly, *Aleurocanthus woglumi* Ash. occurs on citrus, coffee and mango in Central and South America, India, Asia, Malaysia and Indonesia.

As in some species of aphids, the excreta may be copious and sugary and this often results in the growth of black sooty mould fungi on the surface of the plant.

The genus *Bemisia* transmits virus diseases of cotton, tobacco, tomato, sweet potato, cape gooseberry, cassava and okra in Africa. *Bemisia tabaci* (Genn.) one of the main vectors of these diseases, occurs throughou the tropics.

Table 5. Some mealybugs (family—Pseudococcidae) attacking tropical plants

Species	Common name	Hosts	Distribution
Ferrisia virgata (Ckll.)	Guava mealybug, striped mealybug	Polyphagous, including guava, cocoa (vector of swollen shoot virus), coffee, citrus, cotton, jute	Americas, Africa, India, Asia, New Guinea, Pacific, Australia
Dysmicoccus brevipes (Ckll.)	Pineapple mealybug	Pineapple, palms, sugar cane, rice, coffee (vector pineapple wilt and green spot)	Americas, Africa, India, Ceylon, Philippines, Malaysia, Indonesia, New Guinea, Australia, Pacific
Planococcus citri (Risso)	Citrus mealybug, coffee root mealybug	Citrus, cocoa (vector of swollen shoot disease), coffee, polyphagous	Pantropical
Planococcus lilacinus (Ckll.)	Long-tailed mealybug	Cocoa, guava, other fruit and shade trees, citrus, coffee	Madagascar, India, Ceylon, Malaysia, Philippines, Indonesia, Mariana Islands

Pseudococcus adonidum (L.)	Hibiscus mealybug	Citrus, coffee, cocoa, coconut and other palms; greenhouse and ornamental plants	Americas, Africa, Madagascar, India, Ceylon, Indonesia, Australia, Pacific
Maconellicoccus hirsutus (Green)		Hibiscus, Gossypium and other Malvaceae, legumes, tropical fruit and shade trees	Africa, India, South East Asia, New Guinea
Saccharicoccus sacchari (Ckll.)	Pink sugar cane mealy-bug	Sugar cane, sorghum, grasses	Americas, Africa, Madagascar, India, Philippines, Indonesia, New Guinea, Australia, Pacific
Planococcides njalensis (Laing)		Cola, cocoa (vector of swollen shoot disease), coffee	Africa
Nipaecoccus nipae (Mask.)		Sweet potato, potato, avocado, coconut, guava, soursop, etc.	West Indies, Central and South America, India, North Africa, Pakistan, Vietnam

Superfamily 9: Coccoidea (Mealybugs and Scale Insects)

All the members of this superfamily are sap sucking plant parasites. The mealybugs (family: Pseudococcidae) are small insects covered with white, powdery wax. They remain mobile throughout their life. The females are wingless while the males can be wingless or possess a single pair of wings (resembling small flies).

Planococcus citri (Risso) is a polyphagous species occurring throughout the tropics. *Ferrisia virgata* (Ckll.) is almost as widely distributed and attacks a variety of tropical crops. *Planococcoides njalensis* (Laing) is the most important vector of the swollen shoot virus of cocoa in Africa but *P. citri* and *F. virgata* are also vectors of this disease. Some mealybugs of economic importance in the tropics are listed in Table 5.

The scale insects are divided into nine families. The family Eviococcidae resembles the mealybugs in habits, and includes the cochineal insect *Dactylopius coccus* Costa, a native of Central America which feeds on Opuntia and Nopalea. The dried insects are used to produce the crimson cochineal pigment which has long been used in cosmetics and for colouring food and drinks. The evaporated honeydew of *Trabutina mannipara* (Ehr.) on *Tamarix* spp. is the "manna" of Biblical note. It is still used for sugar in parts of the Middle East.

In the remaining families of Coccids only the first instar nymph is mobile and it is in this stage that dispersal takes place. The legs become reduced or disappear in the later stage nymphs while the adult females are little more than bags of eggs or young. Many species secrete layers of protective material which covers their bodies. In the family Lacciferidae this is exploited commercially by the culture and collection of *Kerria lacca* (Kerr) for the manufacture of stick lac which is used in shellac manufacture. (See Ch. 1.)

The armoured scales of the family Diaspididae manufacture a hard waxy scale to protect their body. This family includes a number of major economic pests including several species of citrus scales, the San Jose scale of deciduous fruit trees and Parlatoria scales attacking palms. Some twenty-four genera are of economic significance.

The family Margarodidae includes the cottony cushion scale, *Icerya purchasi* Mask., which occurs throughout the tropics on citrus, mango, guava and a variety of other host plants. Members of the genus *Margarodes* are root feeders and some of these can survive very long unfavourable periods by forming cysts and becoming quiescent.

The large family Coccidae includes a number of genera with pest species, e.g. *Saissetia, Coccus, Pulvinaria, Ceroplastes, Ericerus* and *Lecanium.*

Some scale insects of economic importance in the tropics are listed in Table 6. Many of the coccids produce honeydew and this is sometimes

Table 6. Some tropical scale insects of economic significance

Species	Common name	Hosts	Distribution
Antonina graminis (Mask.)	Rhodes grass scale	Sorghum, gramineae	Pantropical
Aonidiella aurantii (Mask.)	California red scale	Citrus, fruit trees, shrubs, flowers	Pantropical
Aonidomytilus albus (Ckll.)		Cassava, solanum, etc.	West Indies, South America, Africa, Madagascar, India
Aspidiella hartii (Ckll.)	Yam scale	Yam, ginger, eddoe, turmeric	Panama, West Indies, West Africa, India, Ceylon, Pacific
Aspidiotus destructor Sign.	Coconut, transparent or bourbon scale	Coconut, mango, banana	Pantropical
Aulacaspis tegalensis (Zehnt.)	Sugar cane scale	Sugar cane	Africa, Malaysia, Java, Philippines
Ceroplastes destructor Newst.	White wax scale	Coffee, citrus, various fruit and shade trees	Africa, Madagascar, New Guinea, Australia
Ceroplastes rubens Mask.	Pink wax scale	Coffee, citrus, tea, fruit trees, etc.	Africa, Asia, Australia, Pacific

contd.

Table 6 (*contd.*)

Species	Common name	Hosts	Distribution
Chrysomphalus dictyospermi (Morg.)	Spanish, red, or dictyospermum scale	Citrus, deciduous fruit trees, palms	Pantropical
Chrysomphalus ficus Ashm.	Florida red scale	Citrus, wide range of hosts	Americas, Africa, India, Philippines, Malaysia, Indonesia, Australia
Coccus hesperidum L.	Brown soft scale	Citrus, tea, fruit trees, etc.	Pantropical
Coccus viridis Green		Coffee, mango	Widespread
Diaspis bromeliae Kern		Pineapple	
Howardia biclavis (Comst.)	Mining scale	Coffee, cocoa, citrus, tropical fruits	Americas, Africa, India, Ceylon, Java, Australia
Icera purchasi Mask.	Cottony cushion scale	Citrus, mango, guava and other plants	Pantropical
Icerya seychellarum (Westw.)		Polyphagous	Africa, Madagascar, India, Ceylon, South East Asia, Pacific

Ischnaspis longirostris Sign.	Black thread scale	Coconut, palms	Central and South America, Africa, Indonesia
Lepidosaphes beckii (Newm.)	Citrus mussel scale, purple scale	Citrus	Pantropical
Lepidosaphes gloverii (Pack.)	Glover's scale, long scale	Citrus	Pantropical
Parlatoria blanchardii (Targ.)	Date palm scale	Date palm and other palms	Brazil, North Africa, Middle East
Parlatoria oleae (Colv.)	Olive scale	Wide range of trees and shrubs, especially olive	South America, Mediterranean, Middle East, India, Pakistan
Pinnaspis buxi (Bch.)		Coconut and other palms, buxus, pandanus	Americas, Africa, Malaya, Philippines, New Guinea, Pacific
Pseudaulacaspis pentagona (Targ.)	White peach scale	Polyphagous, mulberry, peach, apricot and other fruit trees	Pantropical

contd.

Table 6 (*contd.*)

Species	Common name	Hosts	Distribution
Pulvinaria psidii Mask.	Green shield scale, guava mealy scale	Coffee, citrus, mango, guava and others	Pantropical
Quadraspidiotus perniciosus (Comst.)	San Jose scale	Most deciduous fruit trees and shrubs	Widespread—but mainly temperate
Saissetia coffeae (Walk.)		Coffee	
Saissetia oleae (Bern.)	Black scale	Citrus, olive and many other plants	Pantropical
Parasaissetia nigra (Niet.)		Coffee, citrus	
Stictoccus aliberti Vayss		Cocoa	Africa
Stictococcus sjostedti Ckll.		Cocoa	Africa
Unaspis citri (Comst.)	Citrus snow scale white louse scale	Citrus	Americas, Africa, Vietnam, Malaysia, Indonesia, Australia, Pacific
Vinsonia stellifera Westw.		Coconut and oil palms	Africa, India, South America

covered with sooty mould fungi. Many coccids (especially mealybugs) are attended by ants which assist in their dispersal and protect them from predators. This is of considerable importance in the tropics and emphasizes the need for study of the whole ecosystem when pests are being investigated.

Suborder: Heteroptera

The forewings develop as hemelytra with a basally thickened corium and an apical membrane. The wings usually fold flat over the abdomen with their ends overlapping. The insertion of the mouthparts is separated from the prosternum. There are several aquatic groups (Notonectoidea, Ochteroidea) and waterstriders (Gerroidea) which are minor insect predators but are not of economic importance. The aquatic Corixoidea (water boatmen) are mostly phytophagous. Among the terrestrial superfamilies are some valuable insect predators, a few species of medical significance and a large number of important plant pests.

Superfamily: Cimicoidea

Family: Cimicidae (Bedbugs)

These blood-sucking ectoparasites of birds and mammals include two species attacking man: *Cimex lectularius* L., the common cosmopolitan bedbug, and *C. rotundatus* Sign. which occurs in tropical Africa and Asia.

Both species are brown in colour and nocturnal, hiding in cracks in buildings or furniture during the day and coming out at night to feed. Their flattened, wingless bodies are well adapted for such concealment. The eggs are cemented into the adult resting places. The nymphs which resemble the adults in habits pass through five nymphal instars.

When bedbugs are disturbed they emit a characteristic, unpleasant odour. Under tropical conditions the life cycle is completed in 1–2 months and the adults can go without food for several months at a time. They are not disease carriers but have an irritating bite. They can be controlled with DDT or BHC or by fumigation but some insecticide resistant strains have appeared.

Family: Nabidae

This family is a group of predators well developed in the tropics and mainly attacking phytophagous insects and mites. They are therefore beneficial to man. Three species of *Nabis* are important predators of the cotton bollworm *Heliothis zea* (Boddie) in the United States.

Family: Anthocoridae

This family which is closely related to the Cimicidae includes at least one blood-sucking species. However the majority are insect predators, especially

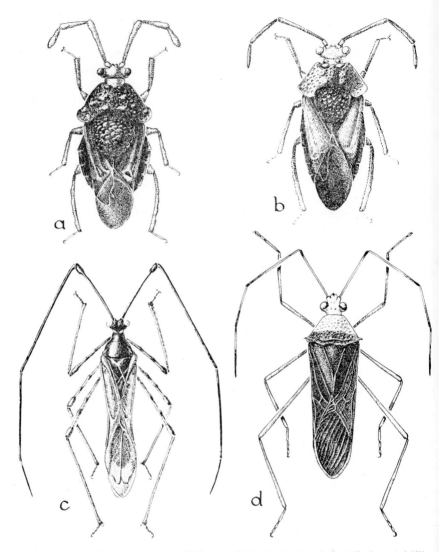

Fig. 4. a. *Pseudodoniella typica* (China and Carv.); b. *Pseudodoniella laensis* Mill.; c. *Helopeltis clavifer* Walk.; d. *Amblypelta theobromae* Brown.

in orchards where species of *Anthocoris* and *Orius* play an important role in biological control of red spider mites as well as attacking aphids, coccids, psyllids, midges and weevils. *Xylocoris flavipes* (Reut.) and *Lyctocoris campestris* (Reut.) are commonly found associated with stored products where they feed on mites and on the larvae of beetles and moths.

Family: Miridae

This family is one of the largest families of Heteroptera and includes some useful predators and some very damaging plant pests. Sixteen species of capsids are found in apple orchards in Britain and most of them are predacious on red spider mites and aphids though some of them occasionally damage fruit by their probing. *Tytthus mundulus* (Breddin) which occurs in Australia and the South West Pacific, feeds on the eggs of Delphacids. It was introduced into Hawaii for the control of the sugar cane leaf hopper *Perkinsiella*.

Fig. 5. *Helopeltis clavifer* Walk. (family: Miridae)
on cocoa pod.

One of the tropical crops most seriously damaged by capsids is cocoa. They damage the fruit and foliage through their toxic saliva and permit the entrance and establishment of weakly parasitic fungi—sometimes leading to death of the tree. In West Africa the main species are *Sahlbergella singularis* Hag. and *Distantiella theobroma* Dist. while *Bryocoropsis laticollis* Schum. and *Odoniella reuteri* occur in the Ivory Coast and Congo. The main cocoa capsids in Central and South America are *Monalonion atratum* Dist. and *M. dissimulatum* Dist. In Papua New Guinea *Helopeltis clavifer* Walk. attacks cocoa as well as other crops while indigenous species of *Pseudodoniella* and *Parabryocoropsis* cause some damage on cocoa. Other genera have been reported attacking cocoa in Sabah and Madagascar (Figs 4 and 5).

Table 7. Some Capsids attacking tropical crops

Species	Common name	Hosts	Distribution
Helopeltis antonii Sign.	Tea mosquito bug	Tea, cinchona, cashew, guava, avocado	India, East Asia, Malaysia, Indonesia
Helopeltis clavifer Walk.		Cocoa, tea, sweet potato, Ficus	New Guinea, Malaysia
Helopeltis schoutedeni Reut.	Cotton mosquito bug	Cotton, castor, mango, sweet potato	Africa
Crytopeltis tenuis Reut.	Tomato suck bug, tobacco capsid	Tomato, tobacco, potato, brinjal	Africa, Asia, New Guinea, Pacific, Australia
Sahlbergella singularis Hag.		Cocoa	Africa
Distantiella theobroma (Dist.)		Cocoa	Africa
Bryocoropsis laticollis Schum.		Cocoa	Africa
Odoniella reuteri Hag.		Cocoa	Africa
Boxiopsis madagascariensis Lav.		Cocoa	Madagascar
Platyngomiriodes apiformis Ghauri		Cocoa	Malaysia
Pseudodoniella spp.		Cocoa	New Guinea
Monalonion spp.		Cocoa	Central and South America

It appears that in many countries indigenous species of capsids move onto cocoa from other host plants—making control more difficult, particularly since small populations of capsids can be very damaging and fugitive. Some capsids attacking a range of economic plants are listed in Table 7. Studies of capsids on African cocoa have emphasized the importance of an understanding of the ecology of the insects for prevention of infestation. For example, it is important to maintain a continuous canopy of cocoa trees to reduce capsid invasion and breeding.

Superfamily: Reduvioidea

Family: Reduviidae (Assassin Bugs)

These are mainly predators on other arthropods. However in the subfamily Triatominae the vectors of Chagas disease of man are found. This unpleasant trypanosomal disease occurs throughout Central and South America and the causal organisms, *Schizotrypanum cruzi* Chagas, have been found occurring naturally in 36 species of Triatominae. The main vectors differ from place to place. These insects bite man but the trypanosomes are not transmitted in the saliva but are deposited in the faeces. Infection occurs when the organisms are introduced into bites, wounds or mucous membranes by scratching. The eyes are a frequent site of infection. *Rhodnius prolixus* Stål, a vector of *S. cruzi* in the northern part of South America, also transmits *Trypanosoma rangeli*, a species harmless to man but pathogenic to the insect vector.

The important vector species live in houses or huts and feed nocturnally on man. The nymphs undergo five moults and take one to two years to reach maturity. Each moult is preceded by a blood meal 6–12 times the weight of the nymph. Some important domestic species are: *Panstrongylus megistus* (Burm.) (the barbeiro of Brazil). *Triatoma infestans* (Klug) (the vinchuca of Argentina) *Rhodnius prolixus* Stål in Venezuela, Colombia and Guatemala. Partially domestic species include: *Triatoma braziliensis* Neiva in North Brazil, *T.sordida* (Stål) in Argentina, Brazil, Bolivia and Paraguay and *T. dimidiata* (Latr.) in Ecuador.

Chagas disease occurs occasionally in Mexico and Central America and some cases are found in Colombia, Ecuador and Peru. It is common in other parts of South America and infection reaches 10–20% of the population in parts of Brazil, Bolivia, Chile and Argentina. 1,000,000 cases have been estimated to occur in Venezuela alone and this disease is a principal cause of myocarditis there since the heart muscle is frequently damaged. Adults have some degree of resistance to infection and the disease is most serious in children and where nutrition is poor and general health is not good.

As with many tropical diseases effective control requires a broad socio-economic approach to raise the general standard of health through improved nutrition and housing conditions and the removal of vectors by rebuilding houses or the use of insecticidal sprays. BHC or Dieldrin have proved

effective sprays for treatment of daytime resting places of the vectors. It i
important to avoid contamination of bites or mucous membranes by insec
faeces carried on the hands.

Superfamily: Coreoidea

Family: Coreidae (Squash Bugs, Leaf-footed Bugs)

This plant feeding family includes several tropical pests including *Leptocoris*
acuta (Thurs.) which attacks rice in India, Asia, New Guinea and the Pacific
Eggs are laid on the upper leaves of flowering rice plants and the nymphs and
adults feed on the developing grains. This sometimes results in 25–50% los
of the crop. Other species occur in Africa and elsewhere.

Fig. 6. *Dysdercus angularis* F. (family: Pyrrhocoridae)—
the cotton stainer.

The leaf-footed bug, *Leptoglossus australis* (F.), which occurs in Papua New Guinea, the Pacific, Asia and Africa causes severe damage to passion fruit and curcurbits as well as producing fruit-fall in citrus through its feeding activities. It also attacks a wide range of other host plants including legumes.

Fig. 7. *Leptoglossus australis* (F.) (family: Coreidae)—
the leaf-footed bug.

Among the most serious Coreid pests in the South Pacific are *Amblypelta* spp. In Papua New Guinea *A. theobromae* Brown (Fig. 8) is the main one of five species which feed on cocoa pods causing damage similar to that caused by capsids. *A. coccophaga* China attacks cocoa similarly in the Solomon Islands and where attack is severe there may be some twig dieback. This species also causes premature nut fall in coconuts. In Papua New Guinea *A. lutescens papuensis* Brown has a wide host range. It causes premature nutfall in coconuts, stem cracking of papaya and cassava and severe damage to rubber seedlings. It may be controlled on cocoa pods by BHC dusts or 0·15% dieldrin spray. The ant *Oecophylla smaragdina* F. attack *Amblypelta* and keeps it in check.

Fig. 8. *Amblypelta theobromae* Brown (family: Coreidae) attacking cocoa pod.

Superfamily: Lygaeoidea

Family: Lygaeidae (Chinch Bugs)

These are smaller and often brighter coloured than the Coreids. Cotton is attacked by *Oxycarenus* spp. in various tropical countries and by *Oncopeltus sordidus* (Dall.) in Queensland. These seed feeders may cause significant losses. In the United States and the West Indies the chinch bug, *Blissus leucopterus* (Say), causes severe damage to cereals and grasses.

Some predacious members of the family are useful in biological control, e.g. *Geocoris* spp. which feed on red spider mites and mealybugs. *Geocoris pallens* Stål is an effective predator of the cotton bollworm *Heliothis zea* (Boddie) in California. *Germalus pacificus* Kirk feeds on the eggs of the fruit fly *Dacus passiflorae* Frogg. in Fiji. However it also feeds on *Teleonemia antanae* Dist. (family: Tingidae) which is an insect used for biological control of the lantana plant.

Family: Pyrrhocoridae

These medium sized to large, red and black bugs are quite conspicuous. They are mostly phytophagous and many of them are seed feeders. The most important of these are the cotton stainers of the genus *Dysdercus* (see Table 8).

Table 8. Some cotton stainers (Dysdercus) on tropical crops

Species	Common name	Hosts	Distribution
Dysdercus fasciatus Sign		Cotton, hibiscus, adansonia, abutilon	Africa
Dysdercus sidae Montr.		Cotton, hibuscis, kapok	Australia, New Guinea, Pacific
Dysdercus singulatus (F.)	Red cotton bug	Cotton, hibiscus, kapok	India, Asia, Australia, New Guinea, Pacific
Dysdercus superstitiosus (F.)	Spotted stainer bug	Cotton	Sudan

They damage cotton in several ways. Their feeding on the young fruit (bolls) may damage some of the seeds and also introduces fungus spores which contaminate the lint and damage the seed as well. Thus direct feeding on the seed reduces germination and the yield of oil while *Nematospora* fungus infection of older bolls results in staining and spoilage of the lint (hence the name "cotton stainers"). There are some predacious Pyrrhocorids but they appear to be of minor importance.

Superfamily: Pentatomoidea (Shield Bugs, Stink Bugs)

This is one of the largest groups of Heteroptera and comprises mainly phytophagous forms though members of the subfamily Asopinae (family:

Pentatomidae) are carnivorous. These include *Cantheconidea furcellata* Wolf which attacks the Zygaenid coconut moth *Artona catoxantha* Hamp., the widely distributed *Zicrona coerulea* L. which preys on flea beetles and other insects, and *Oechalia* spp. which prey on the vine moth and fig leaf beetle in Australia. The saliva of these bugs rapidly paralyses their prey.

Some of the family Pentatomidae are very damaging to plants. One of the best known is the widely distributed *Nezara viridula* (L.), the green vegetable bug, which has a wide host range but is particularly a pest on tomatoes, bean and other vegetables. The green potato bug *Cuspicona simplex* Walk. attacks a range of crops in Australia. In New Guinea *Axiagastus cambelli* Dist. feeds on the spathes and young flowers of coconuts causing some loss of production *Antestiopsis spp.* are important pests of coffee in Africa. Eggs are laid on the leaves and the nymphs feed on leaves, young stems and fruits. The green berries often then become infected with a fungus and rot.

The family Tessaratomidae includes the Australian bronze orange bug *Musgraveia sulciventris* (Stal), a pest of citrus which when disturbed emits a jet of caustic fluid which can cause severe skin irritation or blindness.

The family Scutelleridea includes *Tectoris diophthalmus* (Thunb.), the harlequin bug of cotton in Queensland.

FURTHER READING

Entwistle, P. F. (Ed.) (1964). *Proceedings of the conference on Mirids and other pest. of cocoa at the West African Cocoa Research Institute, Nigeria 24–27 Mar. 1964* Ibadan, pp. 132.

Kennedy, J. S., Day, M. F. and Eastop, V. F. (1962). *A conspectus of aphids as vector of plant viruses.* Commonwealth Institute of Entomology, London, pp. 114.

Kennedy, J. S. and Stroyan, H. (1959). Biology of aphids. *Ann. Rev. Ent.*, **4,** 139–160

Miller, N. C. E. (1956). *The Biology of the Heteroptera.* Leonard Hill, London pp. 162.

Poisson, R. and Pesson, P. (1951). Super-ordre des hémiptéroides in Grassé P. (Ed. *Traité de Zoologie*, Vol. 10. Masson et Cie, Paris, pp. 1385–1803.

Schmitz, G. (1958). Helopeltis du contonnier en Afrique centrale. *Inst. Nat. Etude Agron. Congo Belge, Ser. Scient. no. 71*, pp. 177.

Chapter 8

Lepidoptera: Moths and Butterflies

These familiar insects have two pairs of wings and the body, wings and appendages are covered with overlapping scales which are often arranged in brilliantly coloured patterns. The adult mouthparts are greatly modified and are mandibulate in some primitive families but in all others the mandibles are vestigial while the maxillae are modified to form a sucking proboscis which is coiled up at rest. This is one of the four largest insect orders with more than 105,000 known species.

Because of their beauty they have been popular with amateurs for many years and the butterflies in particular are among the best known insects from the systematic point of view. However, the classification of this order does present some difficulties. The old division into moths and butterflies was artificial and today the order is divided into four suborders: Zeugloptera, Dachonympha, Monotrysia and Ditrysia. The first two of these are small and uneconomic and will not be considered further. The Monotrypsia are of interest because of the superfamily Hepialoidea, while the majority of the Lepidoptera belong to the large suborder Ditrysia.

The eggs are normally provided with a tough shell or chorion which may be variously coloured or shaped and eggs may be laid singly or in batches. The larvae are the main feeding stage and in many cases they can be identified to family or even species by studying the distribution of setae and other characters. The larvae are usually elongated and cylindrical and possess three pairs of thoracic legs and quite often 2–5 pairs of prolegs at the posterior end of the abdomen. Some species are provided with poisonous spines or glandular hairs or with glands secreting repellent odours. Others are brightly coloured as a warning to birds and other predators that they are distasteful. They are predominantly phytophagous and mostly feed on the aerial parts of plants. Many are quite destructive and the order includes many major agricultural pests and stored products pests.

The Lepidoptera undergo a complete metamorphosis (Holometabolous) and the transition from the larva to the quite dissimilar winged adult is accomplished in the pupal stage. Pupation takes place in a silken cocoon woven by the last instar larva. The paired labial silk glands are homologous with salivary glands and are greatly developed in the Bombycoidea where they are commercially utilized for silk production.

The evolutionary success of the Lepidoptera has been associated with the development of the higher plants and most parts of the plant are exploited. In view of the great size and diversity of the order it is convenient to consider the feeding patterns of Lepidoptera in general. There are six main kinds of plant damage though some species may cause more than one type. Thus we can recognize:

(a) *Leaf worms* which are generalized leaf feeders and are therefore quite exposed to parasites and predators. Many leaf worms are brightly coloured or are protected by urticating hairs. For the most part these are not major pests unless the leaves are the crop or unless trees are defoliated. Under normal conditions these are usually kept under control by parasites and predators.

(b) *Army worms*. These are leaf worms with an unstable natural control perhaps through the coincidence of appropriate climatic and environmental factors. More study is needed of these factors. The abrupt appearance of large numbers of caterpillars is usually the consequence of oviposition being concentrated in small areas. Oviposition need not be simultaneous as the existence of a summer diapause coinciding with a dry season may result in the accumulation of the eggs of several generation which will later hatch in large numbers, perhaps at the beginning of the following wet season. On the other hand the "sudden" appearance of large numbers of caterpillars may merely be a consequence of the normal exponential rate of increase of insects living under favourable conditions. In any case, the resulting enormous populations can be most damaging to crop plants before they are controlled by human or natural means.

(c) *Cutworms* are larvae of species which usually spend the day underground and feed at night on the stems of plants, often cutting them off at ground level.

(d) *Stem and leaf borers* may start their life as leaf worms but later bore into the host tissue. This gives them some protection from their natural enemies. The stem borers of rice and other cereals are of great economic importance in the tropics. The leaf rollers and leaf tiers are intermediate between this group and the leaf worms.

(e) The *fruit borers* prefer to feed on the fruits of plants and show varying degrees of specialization. Some feed on the surface of the fruit and on leaves as well while others bore into the fruit and attack the seeds while in the extreme case we have the stored products pests which have lost direct contact with the growing plants and feed only on stored seeds. All gradations can be found. The damage done by this group is often serious and may be complicated by secondary damage by fungi or bacteria.

(f) *The fruit piercing moths* are a small group of adults, mainly of the family Noctuidae, which pierce the ripe fruit to suck the juices—and incidentally allowing the invasion of bacteria and fungi. Several species attack citrus.

Some of the major economic pest Lepidoptera attacking tropical crops are considered below in systematic order.

Order: Lepidoptera

Suborder: Monotrysia

Superfamily: Hepialoidea

In the swift moths of the family Hepialidae, larvae of the genera *Zelotypia* and *Aenetus* tunnel vertically in the stems of living shrubs and trees and sometimes in larger roots. Pastures may be damaged by the feeding activities of larvae of *Oncopera* and *Oxycanus* which tunnel vertically in the soil but emerge at night to feed on plants at the surface. This family is well developed in Australia.

Suborder: Ditrysia

Family: Cossidae

These large, night flying moths, sometimes called the goatmoths or carpenter moths, have long, narrow wings and rather a heavy body. The larvae mostly bore in trees, but occasionally feed on herbaceous plants. *Zeuzera coffeae* (Niet.) attacks coffee, tea and cocoa in New Guinea, India, Malaysia and the Philippines. Eggs are laid in cracks in the bark and the larvae bore into the stem. From time to time frass is ejected to the outside through an opening. This boring weakens the plant causing the leaves to wither and allowing the entry of secondary infections.

Family: Tortricidae

Members of this family attack leaves, fruit and seeds. The codlin moth (*Cydia pomonella* (L.)) is a well-known pest of pome and stone fruits. Several species are leaf rollers. Malvaceous plants are attacked by *Crocidosema plebeiana* Zell. *Cryptophlebia ombrodelta* (Low.) is a pest of *Macadamia* and its larvae also feed on seeds of Acacia and other plants. Another member of this genus (*C. encarpa* Mey.) bores into cocoa pods in Papua New Guinea. *Argyroploce schistaceana* (Sn.) is the white borer of sugar cane in Madagascar and South East Asia.

Family: Psychidae (Bag Moths)

The case-bearing larvae of this family are leaf feeders. They are not host specific but they seldom, if ever, change host plants during development. In recent years they have become increasingly serious pests on coconuts, oil palms and cocoa in South East Asia.

Families: Phyllocnistidae and Gracillariidae

The larval leaf miners of these families attack a number of economic host plants, e.g. *Phyllocnistis citrella* (Staint.) the citrus leaf miner of Africa, India and South East Asia and *Gracillaria theivora* (Wlsm.) the tea leaf miner and leaf roller of Ceylon. *Acrocercops cramerella* Snell is a cocoa pod borer in Indonesia.

Family: Lyonetiidae

Among the leaf miners attacking economic crops are *Bedellia somnulentella* (Zell.), a cosmopolitan pest of sweet potato and *Leucoptera spp.* on coffee in East Africa. *Opogona glycyphaga* Meyr. injures sugar cane and bananas. On cotton in Northern Australia the first two instars of *Bucculatrix gossypii* Turn. are apodous leaf miners but after the third instar the larvae feed on the leaf surface and skeletonize it.

Family: Tineidae

A number of stored products pests are found in this family. Attacking wool, fur and feathers are the case-making clothes moth *Tinaea pellionella* (L.), the common clothes moth *Tineola bisselliella* (Hummel) and the tapestry moth *Trichophaga tapetzella* (L.). Cosmopolitan pests of grain and other foodstuffs include *Nemapogon granella* (L.) and *Niditinea fuscipunctella* (Haw.).

Family: Yponomeutidae

An important and very widespread pest of Cruciferae is the diamond back moth *Plutella xylostella* (L.) which skeletonizes the leaves. When disturbed the larvae drop from the plant on silken threads. The olive leaf miner *Prays oleelus* (F.) also damages olive flowers and fruit in the Mediterranean and South Africa. Two other species of *Prays* attack citrus flowers in Australia and Malaya.

Family: Oecophoridae

Two widely distributed stored products pests belonging to this family are *Hofmannophila pseudospretella* (Staint.) and *Endrosis sarcitrella* (Stephn.) *Hofmanophila* feeds on stored bulbs and tubers and other stored products while *Endrosis*, also a domestic pest, feeds on stored fruits and cereals.

Family: Xyloryctidae

The coconut leaf feeding caterpillar, *Nephantis serinopa* Meyr. is sometimes a serious pest in India, Ceylon, East Pakistan and Burma. *Cryptophasa* spp. feed on the bark of Acacia, Cassia, Citrus, Casuarina and rain trees, sometimes ringbarking the stem. Their tunnels are covered with a web of silk. The cocoa bark web worm *Pansepta teleturga* Meyr. is of increasing importance in Papua New Guinea. Early instars feed on the bark surface but from the third instar they bore deeply into the branch, weakening it and sometimes causing defoliation or death.

Family: Gelechiidae

A very important pest of cotton throughout the Tropics is the pink bollworm, *Pectinophora gossypiella* (Saunders) which attacks the fruit of cotton, Hibiscus and other Malvaceae. Similar damage is caused by *P. scutigera* (Hold.) in Australia and New Guinea; by *P. malvella* Hb. in North Africa and South East Asia; by *P. vilella* Zell. in South Europe and in the Soviet Union and *P. erebidona* Mey. in Uganda. *Phthorimaea operculella* (Zell.) is a widespread leaf miner in tobacco and potatoes and also tunnels in stored potato tubers. This species is commonly known as the potato tuber moth. Other Solanaceous pests are *Scrobipalpa heliopa* (Low.) the tobacco stem borer, which attacks cultivated and wild tobacco and eggplant in Africa, India, South East Asia, New Guinea and Australia, and *S. plaesiosema* (Turn.) which tunnels in the stems of tomatoes.

The larvae of the widely distributed Angoumois grain moth, *Sitotraga cerealella* (Oliv.), are serious pests of stored grain, cereals, wheat, maize, pods, rice, sorghum, millet, buckwheat, cocoa beans and sweet chestnuts. Severely attacked cereals have an unpleasant taste and cannot even be used as animal feed.

Family: Zygaenidae

The coconut leaf moth *Artona catoxantha* (Hamp.) attacks coconut and other palms in Malaysia, Indonesia and West Irian.

Family: Limacodidae

The larvae of the cup moths have greatly reduced head and legs and are slug like in appearance. They are often brightly coloured and bear tufts of ever sible stinging hairs. Apart from direct damage caused by feeding on foliage o fruit they are a considerable nuisance to tea or coffee pickers because of th intense skin irritation resulting from contact with their urticating hairs Pupation takes place in oval or pear-shaped cocoons with a lid. The appear ance of empty cocoons cemented to the leaf has given rise to the name "cup moths".

Setora nitens (Wlk.) feeds on tea, coffee and coconuts in South East Asia and West Irian. *Parasa vivida* Walk. sometimes interferes with coffee harvest ing in Africa, India and Ceylon. *Parasa* and *Thosea* spp. are similar pests o tea while *Parasa pallida* Bosch. feeds on palm leaves in India, Ceylon and Indonesia. *Thosea sinensis* Walk. has been reported defoliating pepper in China, India and Indonesia. Bananas and hemp have been defoliated by *Scopelodes dinawa* Bak. in Papua, New Guinea. Other host plants of cup moths include cocoa, oil palm and mango.

Family: Hyblaeidae

The sole genus in this family, *Hyblaea* was formerly included with the Noctuidae. *H.puera* Cram. feeds on plants of the family Verbenaceae and has been reported defoliating teak in Java and New Guinea. It occurs throughou South East Asia.

Family: Pyralidae

This large family of small or medium sized moths includes some species o great economic importance—particularly the stem borers of rice, sugar cane and maize. There are also a number of leaf feeders, flower, and fruit feeders and pests of stored products.

Although eighteen species of Pyralids have been recorded as rice stem borers (together with *Sesamia* spp. of the family Noctuidae), only four or five of these are widespread and of major economic importance. These are the yellow borer, *Tryporyza incertulas* (Wlk.) of India and South East Asia and the striped stalk borer *Chilo suppressalis* (Wlk.) which has a similar distribu tion but extends to New Guinea. The white borer *Tryporyza innotata* (Wlk.) is the most serious pest in Malaysia, Indonesia and the Philippines. *Chilo auricilius* (Dudg.)—a major sugar cane pest—is of localized importance in India and Taiwan while *Chilo polychrysa* (Meyr.) is important in Malaysia. *Chilo partellus* (Swin.) which usually attacks maize and juar, also attacks rice in India and East Africa. In Central and South America the lesser cornstalk or rice borer *Elasmopalpus lignosellus* (Zeller) attacks rice. The Pyralid stem borers of rice, maize and sugar cane are summarized in Table 9.

Fig. 9. *Nacoleia octasema* (Meyr.) (family: Pyralidae). Banana scab moth.

The eggs of most rice stem borers are laid on the leaves and the newly hatched larvae bore into the stem where they feed throughout their life. Plants are often killed or fail to produce normal seeds resulting in white ears or dried up inflorescences. These insects are amongst the most destructive of agricultural pests and there is need for further study on their ecology and behaviour. The corn and sugar cane borers frequently feed on leaves and inflorescences as well as boring in the stem.

A number of other Pyralids are leaf feeders and in the subfamily Phycitinae there are several well known cosmopolitan stored products pests (*Ephestia, Cadra, Plodia*) as well as the pantropical legume pod borer *Etiella zinckenella* (Treit.). A famous member of this subfamily is *Cactoblastis cactorum* (Berg.) from Argentina which was introduced for the biological control of prickly pear in Australia.

Some of the important tropical Pyralids are summarized in Tables 9 and 10.

Family: Hesperiidae (Skippers)

The small, dull coloured butterflies of this family include two species attacking legumes. These are *Rhopalocampta forestana* feeding on the leaves of *Canavalia*

Table 9. Some major Pyralid stem borers of rice, maize and sugar cane

Species	Common name	Type of borer	Other hosts	Distribution
Tryporyza incertulas (Wlk.)	Yellow or paddy borer	Rice		India, South East Asia
Tryporyza innotata (Wlk.)	White borer	Rice		Malaysia, Indonesia, Philippines, New Guinea, Australia
Chilo suppressalis (Wlk.)	Striped stalk borer	Rice	Kibi, Ikri	India, South East Asia, Australia
Chilo partellus (Swin.)	Maize and juar borer	Rice, maize, sugar cane	Sorghum, juar, ragi, marua, baira	India, East Africa
Chilo auricilius Dudg.	Sugar cane stem borer	Rice, sugar cane	Juar	India, Taiwan
Chilo infuscatellus Sn.		Rice, sugar cane		India, Burma, Philippines, Indonesia
Chilo polychrysa (Meyr.)		Rice, maize		Malaysia, Indonesia
Niphadoses spp.	Paddy stem borer	Rice		India, Australia
Diatraea saccharalis (Fab.)	Oriental sugar cane borer small moth borer	Rice, maize, sugar cane	Mainly on sugar	Americas
Diatraea zeacolella (Dyar)	Lesser corn stalk borer	Maize		USA, Mexico

Diatraea sticticraspis (Hamp.)	Indian sugar cane borer	Sugar cane		India
Zeadiatraea lineolata (Wlk.)	Neotropical corn stalk borer	Rice, maize, sugar cane		Central and South America
Proceras indicus Kapur	Internodal borer	Rice, sugar cane	Millet, ikri, rahri	India
Proceras sacchariphagus Bojer	Spotted borer	Sugar cane		Madagascar, Indonesia
Elasmopalpus lignosellus (Zeller)	Lesser corn stalk borer	Rice, maize	Legumes	Central and South America
Maliarpha separatella (Rag.)		Rice, maize, sugar cane		Africa
Eldana saccharina Wlk.		Maize, sugar cane		Cosmopolitan
Emmalocera depressella Swin.	Cane root borer	Sugar cane		India, Indonesia
Ostrinia furnacalis (Gn.)		Maize (also feeds on leaves)		Asia, New Guinea, Australia, Pacific
Ostrinia nubilalis (Hb.)	European corn borer	Maize		Europe, North Africa, USA
Hypsotropa subcostella Hamp.		Rice		Africa

contd.

Table 10. Some tropical Pyralids of economic importance, feeding on leaves, flowers and fruits

Species	Common name	Feeder	Distribution
Nacoleria octasema (Meyr.)	Banana scab moth	Fruit	Indonesia, New Guinea, Pacific
Coniesta ignefusalis Hamp.	Millet borer		Africa
Nymphula depunctalis (Guen.)	Paddy case bearer	Leaves, rice and grasses	Africa, India, South East Asia, Australia, South America
Hyblaea puera Cram.		Leaves, defoliates teak and Verbenaceae	Indonesia, New Guinea
Sylepta derogata (F.)	Cotton leaf roller	Leaves	Africa and Asia
Sylepta balteata (F.)	Tea leaf roller	Leaves, also stem boring in rice	Nigeria, India, Malaysia
Susumia exigua (Butl.)		Leaves, rice	South East Asia, Australia, Pacific
Crocidolomia binotalis Zell.		Leaves, cabbage and other crucifers	Africa, India, South East Asia, New Guinea, Australia, Pacific
Hellula phidilealis Wlk.		Leaves, cabbage and other crucifers	Central and South America

Species	Common name	Food/habit	Distribution
Cactoblastis cactorum (Berg.)		Leaves, opuntia. Used for biological control	South America, Australia
Coleoneura trichogramma		Flowers and young coconut fruit	Fiji
Tirathaba rufivena Wlk.	Coconut spathe moth	Flowers and young coconut fruit	New Guinea, Solomon Islands
Tirathaba fructivora Wlk.		Flowers, oil palms	Malaysia
Orthaga exvinacea Mi.	Mango web worm	Leaves	India
Margaronia spp.	Melon worms (on cucurbits)	Leaves, flowers, fruit	West Indies, North and South America, West Africa
Marasmia trapezalis Guen.		Leaves, sugar cane, grasses, rice	West Indies, South America, Nigeria, Philippines
Dichocrocis crocodera (Meyr.)		Leaves, coffee, leaf roller	West Africa

contd.

Table 11. Some tropical Pyralid pests of seeds and stored products

Species	Common name	Food	Distribution
Etiella zinckenella (Treit.)	Lima bean pod borer	Pods and seeds of legumes	Pantropical
Ephestia kuehniella (Zell.)	Mediterranean flour moth	Milled stored products	Cosmopolitan
Cadra cautella (Wlk.)		Stored cocoa, grain, figs, etc.	Cosmopolitan
Plodia interpunctella (Hb.)	Indian meal moth	Stored products	Cosmopolitan
Corcyra cephalonica (Staint.)	Rice moth	Stored grain	India
Pyralis farinalis (L.)		Stored products	Cosmopolitan
Aglossa caprealis (Hb.)		Stored products	
Galleria mellonella (L.)	Wax moth	Honeycombs, beeswax	Cosmopolitan
Achroia grisella (F.)		Honeycombs, beeswax	
Doloessa viridis (Zell.)	Green rice moth	Store seeds	Ceylon, Malaysia, Indonesia
Maruca testulalis Gey.	Mung moth	Pod borer of lima beans, Sesame, legumes	Widespread in tropics

in West Africa and *Urbanus proteus* (L.), a minor bean pest in the United States, Two species of *Parnara* are rice leaf scalers in the Philippines. Members of several genera are minor cane pests in the West Indies. Each caterpillar folds the cane leaf in a characteristic manner and lines it with silk. Many of the Hesperiids feed on grasses or palms.

Family: Papilionidae (Swallow Tails)

Several species of the genus *Papilio* feed on the leaves of Rutaceae (including *Citrus*). Others attack Umbelliferae but in general they are minor pests.

Family: Amthusiidae

The banana butterfly, *Taenaris miops* Kirs. sometimes defoliates banana trees in Papua New Guinea. *Stichophthalma phidippus* L. has been recorded on coconut palms in Malaysia.

Family: Pieridae (Whites, etc.)

The cabbage white butterflies are pests of temperate regions but *Appias epaphia* is a minor pest of Brassicas in West Africa. *Catopsila crocole flava* (Butl.) is a pest of ornamentals which sometimes defoliates cassias in New Guinea.

Family: Nymphalidae

This large family of brightly coloured butterflies includes some pests. *Acraea acerata* (Hew.) feeds on sweet potato leaves in East Africa and ties the leaves together with silken webbing. *A. bonasia* F. attacks jute in Nigeria while *Brassolis sophorae* L. feeds on the leaves of coconuts in Guyana.

Family: Lyeaenidae (Blues, Coppers, Hairstreaks)

The larvae of some species are carnivorous on aphids or ants. Among the plant pests are *Virachola antalus*, *Lampides boeticus* L. and *Euchrysops malathana* which attack the fruit and seeds of legumes in Africa. Ripening coffee beans are attacked by *Dendorix lorisona* in Ghana and by *Virachola bimaculata* Hew. in Sierra Leone.

Family: Geometridae (Loopers)

The larvae of this large family get their name from their looping movement. The slender larvae often resemble twigs or leaf veins. The Araucaria looper,

Milionia isodoxa (Prout.), attacks the foliage of hoop pine (*Araucaria*) in New Guinea and sometimes large trees are defoliated Fig. 10.

Fig. 10. *Milionia isodoxa* (Prout.) (family: Geometridae).

Family: Saturniidae

Some of the very large tropical moths of this family have been utilized for silk production. *Antheraea* spp. produce Shantung silk, Tasar silk and Muga silk. *Philosamia ricini* (Boisd.) which feeds on castor oil leaves (*Ricinus communis* L.) produces Eri silk in Bengal and Assam and has been introduced into Indonesia and New Guinea for the production of silk.

A pest species is *Synthereta janetta* (White) which feeds on the leaves of *Eucalyptus deglupta* in Australia and New Guinea.

Family: Bombycidae

The Oriental silkworm, *Bombyx mori* (L.), originally an inhabitant of China has been introduced into many parts of the world for silk production.

Family: Sphingidae (Hawk Moths)

The large larvae of this mainly tropical family bear a dorsal horn on the eighth abdominal segment. The adults are very powerful fliers and often hover over flowers while they are feeding.

Agrius convolvuli (L.) (Sphinx moth) attacks taro and can be a serious pest of sweet potatoes which may be completely defoliated. This cosmopolitan species appears in the literature under a variety of generic names including *Sphinx* and *Herse*. *Hippotion celerio* (L.) attacks taro (*Colocasia esculentum*) in New Guinea and the Pacific Islands as well as cotton in East Africa.

Fig. 11. *Hippotion celerio* (L.) (family: Sphingidae).

Cephanodes hylas (L.) feeds on coffee leaves in Africa and Malaysia. The tobacco worm, *Protoparce quinquemaculata* (Haworth), attacks Solanaceous plants in North and South America, Europe and Hawaii; while *P. sexta* (Johansen) attacks these plants in the Americas and West Indies.

Family: Noctuidae

This is the largest family of Lepidoptera and it contains a great number and diversity of agricultural pests. Some of the more important ones are summarized in Tables 12 and 13.

Some of the Noctuids are polyphagous with a wide host range and distribution. Thus *Spodoptera exigua* (Hb.) is almost cosmopolitan on a wide range of plants and is sometimes a serious cotton pest in the Sudan. Other species of *Spodoptera* are also polyphagous but have a more restricted distribution. The majority of Noctuids are leaf feeders and their economic importance fluctuates greatly from place to place and year to year. The exposed feeding site renders them accessible to parasites and predators so that

Table 12. Some tropical Noctuid leaf feeders, army worms and adult feeders

Species	Common name	Leaf feeder	Army worm	Adults piercing fruit	Hosts	Distribution
Achaea catacaloides Gn.		*		*	Cola, coffee, banana	West Africa
Achaea catella Guen.	Castor semilooper	*		*	Castor oil	West Africa
Achaea janata (L.)		*			Sweet potato	New Guinea, Philippines
Alabama irrorata (F.)		*			Bean (*Phaseolus*)	Nigeria
Alabama argillacea (Hb.)	Cotton leaf worm	*			Cotton	Americas
Anomis sabulifera (Gn.)	Jute semilooper	*			Jute	Africa, Asia, New Guinea, Australia
Anticarsia gemmatalis Hb.		*			Legumes	USA, Barbados
Cosmophila flava (F.)		*			Cotton	Africa, Asia, Australia
Spodoptera litura (F.)		*			Polyphagous	Asia, New Guinea, Australasia, Pacific
Spodoptera exempta (Wlk.)	African army worm	*	*		Cereals, grasses	Africa, South East Asia, Philippines, New Guinea, Australia

Species	Common name			Crops/hosts	Distribution
Spodoptera exigua (Hb.)			*	Sugar beet, cotton, lucerne, tobacco, tomato, peas	Africa, Asia, New Guinea, Pacific, Australia
Spodoptera littoralis (Boisd.)	Egyptian cotton worm *			Polyphagous	Africa, Mediterranean
Spodoptera frugiperda (Smith)	Fall army worm		*	Polyphagous	Americas
Spodoptera mauritia (Boisd.)	Paddy swarming caterpillar		*	Rice, maize, sugar cane, grasses	Madagascar, Asia, New Guinea, Australia, Pacific
Tiracola plagiata (Wlk.)	Cocoa army worm		*	Cocoa, banana, cassava, castor, citrus, coffee, rubber, tea, tobacco	Ceylon, South East Asia, New Guinea, Australia, Pacific
Mythimna loreyi (Dup.)			*	Maize, rice, wheat, sorghum, sugar cane	Mediterranean, Africa, Asia, New Guinea, Solomon Islands
Mythimna unipuncta (Haw.)	American army worm *			Cereals, forage crops	Americas, Mediterranean, Africa
Mythimna separata (Wlk.)			*	Cereals, forage crops	South East Asia, New Guinea, Australasia
Mythimna convecta (Wlk.)	Australian army worm *			Cereals and forage crops	Australia
Agrotis segetum (Schiff.)		*		Tobacco, coffee, maize, wattle	Africa
Agrotis ipsilon (Hfn.)	Greasy cutworm	*		Tobacco, lucerne, potato, cotton, crucifers	Cosmopolitan
Chrysodeixas spp. (syn. Plusia)			*	Polyphagous on dicotyledons	Cosmopolitan (especially *C. chalcites* (Esper) (Fig. 13)

contd.

Table 13. Some tropical Noctuidae attacking fruits, flowers or seeds and stemborers

Species	Common name	Larvae on fruits or seeds	Stem-borers	Hosts	Distribution
Diparopsis spp.	Red bollworm	*		Cotton	Africa
Earias biplaga Wlk.		*		Cotton, malvales, cocoa	Africa
Earias vittella (F.)	Northern rough bollworm	*		Cotton, hibiscus	India, South East Asia, New Guinea, North Australia
Earias insulana (Boisd.)	Spiny bollworm	*		Cotton, hibiscus, Abutilon	Mediterranean, Africa, Asia (to West Malaysia)
Heliothis virescens (F.)		*		Cotton, tomato, tobacco	Americas
Heliothis armigera (Hb.)	Corn-ear worm	*		Maize, tobacco, vegetables, cotton	Europe, Africa, Asia, New Guinea, Australasia, Pacific
Heliothis zea (Boddie)	Cotton bollworm, corn-ear worm, etc.	*		Maize, cotton, tomato	Americas

Species	Common name		Host plants	Distribution
Heliothis assulta Gn.	Cape gooseberry budworm	*	Maize, Physalis, tobacco, tomato, other Solanaceae	Africa, India, South East Asia, New Guinea, Pacific, Australia
Heliothis punctigera Wllgr.	Australian budworm	*	Tobacco, flax, lucerne, sunflower, etc.	Australia
Scadodes pyralis Dyar		*	Cotton	South America
Busseola fusca Fuller	Maize stalk borer	*	Maize, cereals	Africa
Sesamia inferens (Wlk.)	Violet rice stem borer	*	Maize, sugar, cereals	India, Asia, New Guinea, Solomon Islands
Sesamia cretica Led.	Durra stem borer	*	Sorghum, maize, millet, wheat, sugar cane	Mediterranean, Africa
S. botanephaga (Tams and Bourke) and *S. calamistis* (Hamp.)		*	Maize, rice, sugar cane	Africa

natural enemies often limit their population growth, and they are seldom serious pests. However, when environmental conditions are appropriate massive populations of *Spodoptera* or *Tiracola* can occur and these "army worms" can be very damaging until natural controls can come into effect. In the case of *Tiracola plagiata* Walk. occasional local infestations occur on cocoa

Fig. 12. *Tiracola plagiata* Walk. (family: Noctuidae).

but in some years these insects appear in plague proportions. Populations build up on weeds or secondary jungle growth and then migrate in large numbers onto cocoa. Some of the shade trees grown with cocoa, such a *Leucena* and *Crotalaria* are favourable hosts for *Tiracola* and assist in the rapid build up of populations. Under severe attack cocoa trees may be defoliated. Papaya and Cassava are also attacked in New Guinea (Fig. 12).

Agrotis is the best known genus of cut worms and these attack a variety of crops in both tropical and temperate regions. *A. ipsilon* (Hfn.) is polyphagous and the adults can fly long distances.

Those species which feed in the larval stage on fruits or developing seeds include the economically very important cotton bollworms (*Earias* spp.) and corn-ear worms (*Heliothis* spp.) *Heliothis* is a very important and damaging pest of maize, cotton and a wide range of other crops. The larvae of *H armigera* (Hb.) large numbers of young, developing cotton buds. The larvae also bore into the fruits of tomato and okra. They destroy the flower buds flowers and pods of beans and other legumes and feed on the developing grain maize and sorghum. They are not easy to control.

Fig. 13. *Chrysodeixas chalcites* (family: Noctuidae).

The stem borers of rice, maize and sugar cane are also serious pests occurring in this family. The violet rice stem borer *Sesamia inferens* (Wlk.) is a serious pest in India and Asia. Other species of *Sesamia* are found in Africa and the Mediterranean area (Table 12). Some Pyralid moths are also important stem borers of these crops.

FURTHER READING

Benson, J. F. (1973). The biology of Lepidoptera infesting stored products. *Biol. Rev.* **48**(1). 1–26.

Hinton, H. E. (1946). On the homology and nomenclature of the setae of the lepidopterous larvae, with some notes on the phylogeny of the Lepidoptera. *Trans. Roy. Ent. Soc. London* **97**. 1–37.

Fracker, S. B. (1915). The classification of lepidopterous larvae (2nd. Ed. 1930). *Illinois Biol. Monogr.* **2**(1). 1–161.

Portier, P. (1949). La biologie des Lepidopteres. *Encycl. Entom.* **23**. 1–643.

Chapter 9　　　　　*The Flies and Fleas*

Although the fleas of the order Siphonaptera belong to the higher insects (Oligoneoptera), the adults never have wings. They are rather an isolated group of ectoparasites with larval stages resembling some of the Diptera. The smooth, laterally flattened body of the adult is well adapted for their ectoparasitic habitat. The legs are powerfully built for jumping and the mouthparts are adapted for piercing the skin and sucking blood.

The eggs, which are laid on or near the host, or on the ground, hatch to give legless larvae with a well developed head. These feed on dust and domestic debris and particularly prefer the partly digested blood found in the excrement of adult fleas. Pupation takes places in a loosely constructed silken cocoon in the larval sites—often in cracks in floors and wallboards.

After emerging the adult seeks out its warm blooded host mammal or bird. The adults of some species at least are very responsive to the vibrations caused by walking or other movement. They can live for long periods (months) without food but after a blood meal the female lays her eggs which hatch in 1–2 weeks depending upon temperature. At moderate temperatures the larval stage takes 10 days to 2 weeks while pupation takes about a further week.

There is a great variation in host specificity; thus some fleas will attack several species of hosts while others are confined to one.

Since the adults are blood suckers they are potential vectors of disease. They are well known for their role in spreading bubonic plague from rats to man. They also transmit murine typhus from rats to man. The species most dangerous to man is the plague flea *Xenopsylla cheopis* (Roths.) which is normally a parasite of rats. If the rat host dies the fleas leaving the body seek out a fresh host. If the rat should be infected with bubonic plague this can be transmitted to man. The food canal of the infected fleas frequently become blocked with plague bacteria (*Pasteurella pestis*). If such a flea subsequently feeds on a human host some of the organisms are regurgitated into the blood of the new host and infection is transmitted. Since bubonic plague is a serious disease of man, a careful watch is kept on all ports to avoid the introduction of rats, especially black rats, which might be infected. This disease was the "black death" which destroyed one third of the population of Europe in the fifteenth century. It still remains a danger in many parts of the world such as Western America and South East Asia where there are reservoirs of disease in wild rodents. The rat flea *Nosopsyllus fasciatus* (Bosc.) is also a vector of plague but seldom bites man.

Three species of fleas commonly cause human annoyance: the human flea *Pulex irritans* (L.), the cat flea *Ctenocephalides felis* (Bouche) and the dog flea *Ctenocephalides canis* (Curtis). Several species of fleas have been implicated in the transmission of endemic murine typhus caused by *Rickettsia mooseri*.

Several other species of fleas are pests of domestic animals, e.g. the stick-tight fleas of the genus *Echidnophaga* which attack poultry and embed their mouthparts firmly into the skin around the head where they remain attached. One species of this genus is capable of transmitting the virus disease myxomatosis of rabbit though it is probably not an important vector in the field.

The chigger, *Tunga penetrans* (L.), which occurs throughout Africa and South America attaches itself firmly to man, especially between the toes and under the toenails where it can be very irritating. The female burrows under the skin of the feet of man or domestic animals, loses its legs and lays eggs inside its burrow.

Fleas may be controlled by improving domestic hygiene to suppress larval breeding sites, by reducing human contact with domestic animals and by the use of insecticidal dusts such as DDT or BHC.

Order: Diptera (Flies)

This large order includes some 75,000 described species amongst which are some of man's greatest enemies. They transmit disease of man and domestic animals, ravage our crops and cause considerable annoyance. On the other hand some are beneficial in reducing carrion and as predators of crop feeding insects.

The Diptera can be divided into three suborders: Nematocera, Brachycera and Cyclorrhapha, which will be considered in turn.

Suborder: Nematocera

The larvae have a well developed head capsule with the mandibles working horizontally. The antennae of the adult are usually longer than the head and thorax and are composed of numerous (6–14) similar segments. The palpi have 3–5 segments. These small, delicate flies have been divided into 18 families but only five of them will be considered here, the first because its larvae are plant pests (gall midges) and the others because the adults are blood suckers of medical or veterinary importance.

Family: Cecidomyidae (Gall Midges)

These tiny flies have highly ornamented antennae and delicate hairy wings. The adults are mounted on slides for systematic examination and should therefore be preserved in alcohol rather than pinned. The larvae of some species are predacious on mites, aphids, coccids and other small insects but the majority are plant feeders and some stimulate the host plant to produce galls which enclose them.

They have a wide range of hosts and amongst the tropical crops serious damage has been reported on sorghum, mangoes, rice, sesame and tomatoes. The sorghum midge *Contarinia sorghicola* (Coq.) occurs wherever sorghum is grown. Larvae infesting the inflorescence prevent seed development and may reduce the yield by 25–50%. The rice midge, *Pachydiplosis oryzae* (Wood Mason), causes a leaf sheath gall ("silver shoot") so that the inflorescence fails to develop. It occurs in Africa, India, Ceylon, Asia, South East Asia and West Irian.

The Nematoceran families with blood-sucking adults are *Culicidae*, *Simuliidae*, *Ceratopogonidae* and *Psychodidae*. The first two families breed in water while the second two breed either in water or damp places. Blood-sucking insects may be pests for several reasons. Their bite may be very irritating though the loss of blood is usually minor. However, their most important role is in the transmission of disease and in this respect the Culicidae are the most important vectors of human disease.

Family: Culicidae (Mosquitoes)

The larvae live in water and the adults have a prominent proboscis directed forwards and possess straight, rigid palps. Scales are present on the body, wings and legs. Although over 2,000 species have been described, only a small number are of medical significance. The family is divided into two sub-families as follows.

Subfamily: Toxorhynchitinae

These do not feed on blood and are not of medical importance. They are large, highly coloured day fliers and may be distinguished by their peculiarly shaped proboscis which is hook-shaped at rest. The larvae breed in small confined collections of water and some are predacious on other species of mosquito larvae.

Subfamily: Culicinae

Tribe: Culicini

The adult scutellum is trilobed with bristles on each lobe and bare areas between them. The abdomen is completely covered with broad scales which nearly always lie flat. The larvae obtain air through a prominent siphon bearing one to several hair tufts and usually a well-developed pecten. The eggs are deposited in tight, raft-like masses or are laid singly on the surface of the water, on mud or on the ground. There are no floats on the eggs. At rest the larvae lie at an angle to the surface of the water. In this tribe the male mosquitoes have long, slender palps about equal to the proboscis in length and the antennae bear long, bushy setae.

One of the most important disease vectors among the Culicines is *Aedes aegypti* (L.) a very widespread breeder in man-made containers throughout the tropics. This transmits the viruses of yellow fever in Central and South America and Africa and dengue and the related hemorrhagic fever in South East Asia and the Pacific. The most important aspect of control is to eliminate the breeding sites of the mosquitoes where possible—particularly close to human habitation. Reservoirs of yellow fever infection remain amongst forest monkeys in both Africa and South America but different species of mosquitoes are involved in transmission between monkeys and between monkeys and man. Several species of Culicines transmit viruses causing encephalitis in Asia, the Americas and Australia. Indeed more than 100 species of mosquitoes have been incriminated in the transmission of viruses of one kind or another and at least 40 of the vectors belong to the genus *Aedes* and 20 to *Culex*.

Several species of Culicines transmit the filarial worms *Wucheraria bancrofti* and *Brugia malayi*, but they are less important in this role than the Anophelines. *Mansonia* spp. are important vectors in South East Asia. *Wucheraria* is also transmitted by the swamp breeding *Culex annulirostris* (Skuse) and *C. bitaenorhynchus* (Giles) and by *Aedes kochi* (Don.) which breeds in Pandanus leaf bases. The association of filarial worms with mosquitoes is widespread: indeed 19 species of filarial worms are known to be transmitted by 30 species of mosquitoes.

Tribe: Anophelini

Here the palpi of both sexes are about as long as the proboscis, the scutellum is evenly rounded and the legs are very long and slender without tibial bristles or pulvillae. The abdomen is without scales, at least on the lower surface, and the wings often bear characteristic spotted marks. The male palps are thickened at the tip. The adults normally feed with the head, thorax and abdomen in a straight line at about 45° to the surface on which they are standing and larvae rest parallel to the water surface.

Of the hundreds of described species of *Anopheles* about 80 have been reported to be of medical significance. However, in any particular region the number of medically important species is small, perhaps less than six. By far the greatest medical significance of the Anophelines is their role in the transmission of malaria, a disease which as recently as 15 years ago was causing 250,000,000 human cases a year with more than 2,000,000 deaths. The socio-economic significance of this disease can hardly be overestimated in the tropics.

The most important vector species vary from one area to another, e.g. in Africa the main malaria vectors are *Anopheles gambiae* Giles and *A. funestus* Giles. Like most efficient malaria vectors *A. gambiae* is high anthropophilic i.e. man is vigorously attacked and is a favoured host. These insects breed in man-made sites close to human habitation, e.g. drains, pits and puddles. The larvae can tolerate extremes of water temperature and adult populations are correlated with rainfall. On the other hand *A. funestus* prefers larger bodies of water of a more permanent nature such as swamps, ponds or lakes, the larvae are less tolerant of high temperatures and populations fluctuate less throughout the year.

In the Oriental region, although there is some overlapping, one vector replaces another as we move from India through South East Asia to Australia. *A. culicifacies* Giles is one of the most important malaria vectors wherever it appears and is particularly important in India, Ceylon and Burma but also occurs in West Pakistan and Afghanistan. Preferred breeding sites are fresh water with grassy edges. Dense shade is avoided when ovipositing. *A. sinensis* Wied. is the most important vector in Central and South China. From South China to Indonesia this is replaced in importance by *A. balabacensis* Baisas in the forest and by *A. sundaicus* (Rod.) on the coast. From India to Malaya *A. minimus minimus* Theob. is the major vector. In the Philippines this is replaced by *A. minumus flavirostris*. From Malaysia to Indonesia *A. maculatus* Theob. is important. In Papua New Guinea *A. punctulatus* (Don.), *A. koliensis* (Owen) and *A. farauti* (Lav.) are important while in Australia only *A. farauti* is a significant vector.

Moving westwards from India, *A. stephensi stephensi* List. extends from West India to the Persian Gulf. *A. sacharovi* Favre overlaps from West Pakistan to North Italy, where it is an important vector, and in turn its distribution is overlapped by *A. labranchiae labranchiae* Fall. the main southern European malaria vector (particularly in Sardinia and Spain).

In the Americas the main vectors are *A. albimanus* Wied. and *A. darlingi* Root. *A. albimanus* is particularly important in the Caribbean region and along the Gulf coast of Mexico. It is a domestic species of hot, humid climates and the larvae are found in a variety of sunlit water collections (swamps, large ponds, lakes or rice fields). *A. darlingi* is also a highly domestic species but prefers deeply shaded stream edges and is often found breeding in masses of surface vegetation in still water.

As stated above, the Anophelines are also major vectors of filarial worms.

Family: Psychodidae (Sand Flies, Moth Flies)

These are small, hairy flies whose larvae mostly feed on decomposing organic matter. Members of the genus *Phlebotomus* transmit both virus and protozoal diseases of man. Four species transmit the group c arbovirus of sand fly fever (Papatasi fever) from the Mediterranean, through Asia to Southern China and South East Asia. The insects which establish themselves in ruins or heaps of rubbish are frequent companions to modern war.

Phlebotomids also transmit diseases caused by Protozoa of the genus *Leishmania*, including oriental sore, Kala-azar and espundia. These diseases are widespread through the tropics and reservoirs of infection are found in both wild and domestic animals.

Family: Ceratopogonidae (Sand Flies, Biting Midges)

The adults are blood suckers of vertebrates and are often extremely annoying on beaches or estuaries, or in swamps where large populations are sometimes encountered. The larvae are mostly found in water: some in grassy mats around ponds, springs, creeks or lakes, other in liquid manure or slimy water while some prefer the edges of salt marshes and mangrove swamps.

Some species of *Culicoides* are vectors of filarial worms of stock as well as of the viruses of blue tongue and horse sickness. These diseases of domestic stock which have a high mortality have recently become widespread around the world.

In West Africa members of the genus *Forcipomyia* are reported to be the main pollinators of cocoa trees.

Family: Simuliidae (Blackflies)

Many of these small, dark, hump-backed biting flies require a blood meal for egg maturation. The larvae and pupae of characteristic shape are found in fast flowing rivers and streams. The adults sometimes occur in enormous numbers, especially in forested river valleys, and they may be considerable pests through their biting activities alone.

Some species of *Simulium* are vectors of the filarial worm *Onchocerca volvulus* (Lemk.) which causes onchocerciasis in man and brings about a high incidence of blindness in Central America and in Africa. The main vector species are *Simulium ochraceum* Walk. in Guatemala and *S. damnosum* Theob. in Africa. The disease and its vectors are associated with life in shady situations along the turbulent streams which are preferred larval sites. They are frequently found in gulley forest in savanna country.

Suborder: Brachycera

The adults of this small suborder have an antenna of less than six segments and ending in an arista. The larvae have a reduced, retractile head and mostly live in damp vegetable matter and are predacious. A few live in water and some are plant feeders. The adults are mainly predacious and some of these which attack other insects (families: Asilidae and Bombyliidae) are beneficial, while others which attack man and livestock (family: Tabanidae) are harmful.

Family: Tabanidae (March Flies, Horse Flies)

The females of this family which attack man and livestock often inflict a painful bite. They cut the skin and lap up the flowing blood rather than piercing and sucking like the mosquitoes. Other species can feed on surface films. The males feed on plant secretions. The larvae live in mud or slow flowing streams and eggs are usually placed on vegetation over such sites. Unlike most mosquitoes the adult Tabanids are usually active in bright sunlight.

Tabanids are responsible for the mechanical transmission of anaplasmosis of stock, virus equine infectious anaemia and tularemia of man. The filarial worm, loa loa, which is widespread though tropical Africa, is transmitted by forest tabanids of the genus *Chrysops* and the life cycle involves both man and other Primates.

Family: Stratiomyidae

The elongate, flattened larvae have a thick, leathery cuticle impregnated with calcareous material. Some are aquatic but many live in damp soil or decaying vegetation. *Altermetoponia rubriceps* (Macq.) which breed in large numbers in soil in Queensland is a serious pest of pastures and sugar cane.

Suborder: Cyclorrhapha

This is the largest suborder of flies including over 30 families. The adults mostly have short, stout bodies and antennae with three segments, the third segment bearing a style or arista. The larval head is greatly reduced and forms a conical, membranous anterior segment of the body. Two mouth hooks (mandibles) connect with an internal pharyngeal skeleton and work vertically instead of on the usual horizontal plane. The fourth larval instar and pupa develop within the larval skin of the third instar which hardens to become a puparium (pupal case). The adult emerges by pushing off a cap at one end of the puparium. In the series Schizophora this process is assisted by the presence of a frontal sac on the head which is inflated to push off the end of the pupal case. This series includes a large number of families which may be separated into two groups.

Group 1: Acalyptratae with small or vestigial membranes (calyptrae or squamae) at the base of the wings.

Group 2: Calyptratae with well developed squamae at the base of the wings.

The Acalyptrates include a large number of families including a few of economic significance, e.g. Trypetidae, Agromyzidae, Chloropidae and Gasterophilidae while among the Calyptratae are the majority of Diptera in the families Muscidae, Oestridae, Calliphoridae and Tachinidae which are of considerable medical and veterinary importance.

From the agricultural point of view the family Trypetidae is of great importance since it includes a number of species of fruit flies which are major pests of fruit. The adults have marbled wings and the phytophagous larvae live in stems, leaves, and particularly in fruit. The most important genus in the Pacific is *Dacus*. The larvae of *Dacus dorsalis* Hend., the Oriental fruit fly which also occurs in Pakistan, India and South East Asia, and of *D. tryoni* (Froggatt), the Queensland fruit fly, feed on a wide variety of fleshy fruits. Eggs are laid on the skin of the fruit and larvae feeding inside the fruit induce it to rot and ruin it commercially. *Dacus musae* (Tryon) is a pest of bananas in Australia and the South Pacific.

Several species attack cucurbits: *Dacus cucurbitae* Coq. the melon fruit fly in India, South East Asia and the Pacific; *Myiopardalis pardalina* (Big.), the Baluchistan melon fly, from Cyprus and Turkey to India; *Dacus ciliatus* Lw., the lesser melon fly which occurs in Africa, the Middle East, Pakistan and India; and *Dacus vertebratus* Bez. which occurs in Africa in higher rainfall areas.

The peach fruit fly of India, *Dacus zonatus* Saund., has a wide host range, while in Central and South America and the West Indies species of the genus *Anastrepha* attack a wide range of tropical fruits. The mediterranean fruit fly, *Ceratitis capitata* (Wied.) attack deciduous and subtropical fruits (especially peach and citrus) around the Mediterranean, in Africa, Madagascar, Central and South America.

Leaf miners of the family Agromyzidae attack a wide range of hosts plants but are seldom of major economic importance. The French bean fly *Melanoagromyza phaseoli* (Try.) is widely distributed in Australia, Asia and Africa in legumes. The larvae mine the leaves, and later the stems and may kill young plants.

In the family Chloropidae (fruit flies and eye flies) several species of *Oscinella* attack rice and cereals in Malaysia and East Pakistan. Some are stem borers while others produce galls on the leaves of rice. However, the greatest importance of this family is the role of *Hippelates* spp. in the transmission of human yaws and conjunctivitis. The eye fly of the Oriental region is *Siphunculina funcicula* (Meij.). These flies are attracted to pus, mucus and blood. Very large populations are sometimes encountered. Sandy, well drained soils are favoured breeding sites.

In the group Calyptratae, the family Anthomyidae, includes a number of agricultural pests of relatively minor importance in the tropics. *Atherigona* spp. attack the stems of young rice plants in the Philippines, Malaysia and Indonesia. *A. varia* var. *soceata* Rond., the central shoot fly, attacks young sorghum and wheat plants in Africa producing "dead hearts". *Hylemya cilicrura* (Rond.), the bean seed fly, onion fly or seed onion maggot, is a polyphagous fly particularly attacking maize, onions and crucifers. It is widely distributed—except in South East Asia.

Apart from these agricultural pests the main importance of the Calyptratae lies in the flesh-feeding habits of some larvae or the blood sucking of the adults of a few species. The main families concerned are:

    family: Glossinidae (tsetse flies)
    family: Muscidae (house flies, stable fly, buffalo fly, screw worms)
    family: Gasterophilidae (not flies)
    family: Calliphoridae (blow flies)
    family: Sarcophagidae (blow flies)
    family: Tachinidae
    family: Oestridae (bot flies, gad flies)
    family: Cuterebridae (bot flies).

(One recent classification includes the Oestridae, Hypodermatidae, Cuterebridae, Gasterophilidae, Calliphoridae and Sarcophagidae in the family Tachinidae.)

The house flies, such as the cosmopolitan *Musca domestica* (L.), lay their eggs in decaying plant or animal material which provides food for the larvae. The adults are a menace to health through the mechanical transmission of pathogenic organisms, especially of dysentery. The blow flies of the families Calliphoridae and Sarcophagidae lay their eggs in carrion and play an important role in its disposal. Some species lay eggs in septic wounds or on soiled skin or hair. Thus *Lucilia* (family: Calliphoridae) lays its eggs on soiled wool of the sheep around the tail and the larvae feed on exudates from the irritated skin.

In *Gasterophilus* (family: Gasterophilidae) this habit of oviposition near a natural orifice has become specialized and upon hatching the larvae crawl into the nose or mouth and enter the body of the host, e.g. the horse bot flies: *G. intestinalis* DeG., *G. nasalis* (L.) and *G. haemorrhoidalis* (L.). The damage to the host is usually slight. *Oestrus ovis* L. females deposit newly hatched

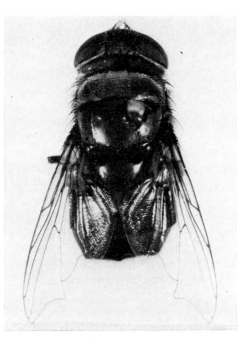

Fig. 14. *Chrysomyia megacephala* (family: Calliphoridae).

larvae around the nostril of sheep. The larvae develop in the nasal sinuses until mature when they are sneezed out to pupate in the soil.

The Congo floor maggot *Auchmeromyia luteola* Fab. is found in huts where the larvae live in cracks in the floor. At night they emerge and pierce the skin of sleepers on the floor and suck their blood. The larvae of another member of the family Calliphoridae also pierce the skin (*Cordylobia anthropophaga* Grun.) but in this case the larvae enter the tissue and cause boil-like swellings.

The larvae of warble flies, *Hypoderma* spp. (family: Hypodermatidae), not only bore into the skin of cattle but also move about inside the body until they reach the back where they form boil-like swellings. The mature larvae escape from a hole in the centre of the swelling and pupate in the soil. The screw worms (family: Calliphoridae) are particularly serious pests of cattle and other domestic livestock. The Old World screw worm, *Chrysomyia bezziana* Vill., occurs in South Africa and Southern Asia extending to New Guinea. The American screw worm, *Callitraga hominivorax* (Coq.), which occurs in the Americas attacks cattle, sheep, goats and pigs and is responsible for annual losses amounting to millions of dollars. Humans can also be attacked. The larvae, hatching from eggs laid in wounds, invade healthy tissue and cause considerable damage. In South East America about 12% of cattle are infested—usually following tick bites. This species aroused considerable scientific interest when populations were eradicated from Curacao, Florida and elsewhere by the use of the sterile male technique. This required the rearing of very large numbers of *Callitraga*, e.g. in 1965 one factory in Texas was producing 100 million sterile adults per week.

In the family Cuterebridae bot flies of the genus *Dermatobia* are widely distributed in the tropics of the New World. They lay their eggs on blood sucking arthropods or on house flies and the eggs hatch and larvae leave their carrier when it alights on the skin of a mammal (including man). They then burrow into the skin, wander about, and eventually form a pocket beneath the skin.

Blood sucking adults are important for their nuisance value to man and livestock and also as vectors of disease. In the subfamily Stomoxinae (family: Muscidae), the cosmopolitan stable fly *Stomoxys calcitrans* (L.) bites man and domestic animals (especially horses). Similarly the buffalo fly *Haematobia exigua* (de Meij.) is a major pest of cattle and horses in the Oriental and Australian regions.

By far the most serious pests among the higher Diptera are the tsetse flies *Glossina* spp. (family: Glossinidae) which are vectors of the trypanosomes responsible for human sleeping sickness and nagana of cattle in Africa. These flies have influenced the migrations of man and the agricultural development of large areas of Africa. Human and animal trypanosomiases are major problems in Africa from 15° North to 20° South. Some 11 species of Trypanosoma are transmitted by 19 species of Glossina. The species responsible for human sleeping sickness, *Trypanosoma gambiense* Dutton, and more rarely *T. rhodesiense* S. and F., are mainly transmitted by *Glossina palpalis* (Robb.-Des.) and *G. morsitans* Westw. respectively, and these are found in different types of habitats (*G. morsitans* in the bush and *G. palpalis* near rivers). The main species of *Trypanosoma* attacking domestic animals are *T. brucei* P. and B., *T. vivax* Ziem. and *T. congolense* Broden. These are all transmitted in the saliva of *Glossina* following multiplication in the alimentary canal.

Some of the Tachinidae are insect parasites and are useful in biological control. One of the Sarcophagidae, *Sarcophaga destructor* Hall, is a pest of the ripe fruits of cucurbits, tomatoes, egg plants and onion bulbs in the Sudan.

FURTHER READING

Barnes, H. F. (1940–56). *Gall midges of economic importance*. Vols. 1–7. Crosby Lockwood, London.

Brauns, A. (1954). *Puppen terricoler Dipterenlarven*. Musterschmidt, Göttingen, pp. 156.

Buxton, P. A. (1955). *The natural history of tsetse flies*. London, pp. 816.

Christophers, S. R. (1960). *Aedes aegypti (L.) the yellow fever mosquito*.

Cole, F. R. (1969). *The flies of western north America*. University of California Press, Los Angeles, pp. 693.

Curran, C. H. (1934). *The families and genera of North American Diptera*. Tripp, N.Y., pp. 512.

Gillett, J. D. (1971). *Mosquitoes*. Weidenfeld and Nicholson, London, pp. 274.

Holland, G. P. (1964). Evolution, classification and host relationships of Siphonaptera. *Ann. Rev. Ent.* **9**. 123–146.

Hopkins, G. H. E. and Rothschild, M. (1953–66). *An illustrated catalogue of the Rothschild collection of fleas (Siphonaptera) in the British Museum (Natural History) Vols. 1–4*. London.

Hubbard, C. A. (1947). *Fleas of Western North America, their relation to public health*. Iowa, pp. 533.

Lindner, E. (1924). *Der Fliegen der Palaearktischen Region*. 14 vols., Schweizerbart'-sche Verlag, Stuttgart.

Marringly, P. F. (1969). *The biology of mosquito-borne disease*. Allen and Unwin, London, pp. 184.

Nash, T. A. M. (1969). *Africa's bane, the tsetse fly*. London, pp. 224.

Nijveldt, W. (1969). *Gall midges of economic importance*, Vol. 8. Crosby Lockwood, London, pp. 220.

Oldroyd, H. (1964). *A natural history of flies*. Weidenfeld and Nicolson, London.

Roberts, F. H. S. *Insects affecting livestock with special reference to important species occurring in Australia*. Angus and Robertson, Sydney, pp. 267.

Rothschild, M. and Clay, T. (1952). *Fleas, flukes and cuckoos*. Collins, London, pp. 304.

Smart, J. (1965). *Insects of medical importance*. 4th edition London, pp. 295.

Zumpt, F. (1965). *Myiasis*. Butterworths, London, pp. 267.

Chapter 10   *Beetles*

This enormous group of insects (order: Coleoptera) includes more species than any other group of organisms. About 300,000 species have been de scribed, comprising some 40% of the known species of insects. They vary enormously in size, structure and habitat. The possession of horny elytra (wing covers) is characteristic; the mouthparts are usually biting and the prothorax is large and mobile. Many economic insects belong to this order and the more important tropical species will be mentioned below under their families.

Modern classification divides the beetles into four suborders. The Arch ostemata and Myxophaga are of no economic significance. The suborder Adephaga comprising one superfamily, the Caraboidea, includes two families of terrestrial forms and five families of aquatic beetles. These are mostly predacious and are of biological interest for their role in food chains. The fourth suborder, the Polyphaga includes an enormous assemblage of beetles and only some of the more important families or superfamilies will be considered.

Although some species produce corrosive chemicals as defence mechanisms, the beetles are of little medical or veterinary importance; however, their agricultural importance is vast. They function in pollination, destruction of weeds, burying of excreta and carcases, soil fertility, reduction of fly breeding sites, predation on harmful insects as well as their negative role in causing extensive injury to crops and stored products.

Suborder: Polyphaga

Superfamily: Scarabaeoidea

Family: Scarabaeidae

The larvae of this large and important family usually feed on roots, dung or decaying vegetable matter while the moderate to large sized adults are often active fliers and feed on living plants.

The subfamily Dynastinae includes some major pests. The larvae feed on decayed matter or roots in the soil, while the adults are active at night and are attracted by lights. Often the adult male bears horn-like structures on the head and pronotum. The rhinoceros beetle, *Oryctes rhinoceros* (L.), is one of the most serious pests of coconuts and other palms in India, Ceylon, South East Asia and the Pacific. The adults bore into the unfolded leaves of the palms, producing a characteristic damage pattern when the leaves unfold (Figs 15 and 16).

Similar damage to palms is caused by *Oryctes monoceros* (Ol.) in Africa as well as by *O. boas* (F.) and *Dynastes centaurus* (F.) in the Sudan. The damage caused by *Oryctes* often permits the very destructive black palm weevil

Fig. 15. *Oryctes rhinoceros* (L.) (family: Scarabaeidae, subfamily: Dynastinae) the coconut rhinoceros beetle.

(*Rhynchophorus bilineatus* (Montr.)) to gain entry to the tree. In New Guinea similar damage to Oryctes is caused by *Scapanes australis* (Boisd.) though this species prefers younger palms (Fig. 17). The widespread and spectacular elephant beetle *Xylotrupes gideon* (L.) (Fig. 18) causes similar damage but is less important economically. Bananas are also damaged by adults of *Scapanes* and *Papana* boring into the pseudostem. *Papuana* spp. also attack vegetables in New Guinea.

Both adults and larvae of the black beetle, *Heteronychus arator* (F.) attack the roots of pasture plants, sugar cane, maize, tobacco, vegetables, etc. in

Fig. 16. Coconut palm fronds damaged by rhinoceros beetle.

Southern Africa, Madagascar and Australia. *Heteroligus meles* Bilb. and *H. appius* Burm. are important larval borers of yam tubers in West Africa. So are *Prionyctes* spp.

In Central and South America, *Strategus jugurtha* Burm., attacks pineapple stems, while several species of this genus are pests of sugar cane and coconuts.

Fig. 17. *Scapanes australis* (Boisd.) (family: Scarabaeidae, Dynastinae).

Subfamily: Melonthinae (Cockchafers)

The C-shaped larvae, which live in the soil, feed on roots and vegetable matter and do a lot of damage to grasses and sugar cane roots. Thus *Dermolepida albohirtum* Waterh. is the most serious pest of sugar cane in north Queensland. Some root-feeding tropical Melolonthids are summarized in Table 14 overleaf.

Fig. 18. *Xylotrupes gideon* (L.) (family: Scarabaeidae, Dynastinae).

Table 14. Some tropical Melolonthids on economic crops

Species	Crops affected	Locality
Cochliotus sp.	Sugar	Tanganyika
Ctenora smithi	Sugar	Barbados
Dermolepida albohirtum Waterh.	Sugar	Queensland, New Guinea
D. nigrum (*Nonf.*)	Banana	New Guinea
Eutheola fugiceps Lec.	Sugar	Southern USA
Exopholis hypoleuca Wied.	Tea	Indonesia
Lachnosterna spp.	Sugar	W. Indies, S. America
Pseudoholophylla sp.	Sugar	Queensland
Psilopholis grandis Cast.	Rubber	Malaysia, Indonesia
Schizonycha cibrata Blan.	Sorghum, groundnuts	Sudan

Subfamily: Rutelinae

The polyphagous rose beetle, *Adoretus versutus* Har., attacks the roots and shoots of cocoa, coffee and other hosts in Madagascar, India, Pakistan, Java and the Pacific. Other species of *Adoretus* attack maize and tobacco in Africa. The adults of *Anomala superflua* Ar. feed on young tea leaves in Ceylon.

Superfamily: Buprestoidea

Family: Buprestidae

The brilliantly coloured adults of this family are abundant in the humid tropics. The larvae feed in the roots or stems of trees or herbaceous plants. The prothorax of the larvae is very broad and they usually gnaw rather flattened galleries beneath the bark. *Chrysochroa bicolor* F. is a cocoa stem borer in South East Asia. Several species of *Sphenoptera* bore in cotton stems in India and Africa. *Agriculus* spp. attack citrus in the Philippines and cowpea in Africa.

Superfamily: Elateroidea

Family: Elateridae

The phytophagous larvae of many members of this family are commonly known as wireworms and feed on the roots of cereals, grasses and root crops. The larvae of *Diasterias*, *Heteroderes* and *Agriotes*, are widespread pests attacking tobacco roots and other plants. *Lacon variabilis* Cand. is a pest of sugar cane in Queensland. The larvae of some Elaterids are predacious.

Superfamily: Dermestoidea

Family: Dermestidae

The larvae of several genera of this family are widely distributed stored products pests feeding on hair, hides, fur, skins, feathers and woollen products. The hide beetle, *Dermestes maculatus* DeG., is responsible for considerable annual losses of baled hides. Both larvae and adults of the bacon beetle, *Dermests lardarius* L., will feed on a wide range of dry or decomposing animal material. The larvae of the smaller *Attagenus*, *Anthrenus* and *Anthrenoceros* are known as carpet beetles because of their feeding habits but they also attack a wide range of dry animal material (including insect collections).

The adults are mostly pollen feeders. Some Dermestids can also feed on dried plant material, e.g. *Trogoderma granarium* Everts, the Khapra beetle, which is a serious pest of stored cereals. Originally Indo-Malayan in distribution, this species has now become established in Europe, USSR, China, Japan, the Philippines, Australia and Madagascar.

Superfamily: Bostrychoidea

The larvae of these small to medium sized beetles feed on wood or dry vegetable or animal materials. They include some important wood borers.

Family: Anobiidae

The cosmopolitan furniture beetle, *Anobium punctatum* DeG., is a well known domestic pest. One of the most important pests of stored tobacco, the tobacco beetle, *Lasioderma serricorne* F., also attacks other dried plant materials such as drugs or peppers. Another cosmopolitan pest of drugs is *Stegobium paniceum* F. which attacks an even wider range of plant and animal products which are stored under dry conditions. The death watch beetle, *Xestobium rufovillosum* DeG., is a less widely distributed timber borer than Anobium.

Family: Ptinidae

This includes some minor household and stored products pests belonging to the genera *Ptinus*, *Niptus*, *Gibbium* and *Mezium*. These small beetles feed mainly on dried plant and animal materials but are usually of little economic significance.

Family: Bostrychidae

Many members of this family infest diseased or newly felled trees. However *Apate monachus* F. is a stem borer of coffee, cocoa and other host plants in Africa and the West Indies. A widely distributed pest of bamboos is *Dinoderus minutus* F. and this also does quite a lot of damage to bamboo furniture. *Rhizopertha dominica* Steph. is a cosmopolitan tropical pest of grain, which is now the most important pest of stored grain in Australia.

Family: Lyctidae

Both adults and larvae of this family feed in dead, dry wood and dry roots of herbaceous plants. The powder post beetle, *Lyctus brunneus* Steph., infests the unseasoned sapwood of hardwoods leaving a powder filled shell. It is cosmopolitan through the tropics.

Superfamily: Cleroidea

Family: Trogossitidae

The cosmopolitan stored products pest *Tenebroides mauretanicus* L. feeds on cereals, cereal products and dried fruit. Its larvae are to some degree predacious on other store products pests. The main food of the cadelle is the embryo and softer parts of seeds and it can be an important pest of stored grain.

Family: Cleridae

The copra beetle, *Necrobia rufipes* DeG., is cosmopolitan in distribution and feeds on oily or fatty materials such as copra, ham or cheese. It is also predacious and is frequently found on dead animals.

Superfamily: Cucujoidea

Family: Nitidulidae

Many of this large family of sap or juice feeding beetles are found associated with damaged or stored fruit but they are seldom responsible for primary plant damage. They appear to be attracted to fermenting fruit. The dried fruit beetle, *Carpophilus hemipterus* (L.), is cosmopolitan and is widely associated with ripe and decomposing fruit in the field. It occurs in a wide variety of dried fruit and other store products and has also been reported damaging both live and stored figs.

Family: Silvanidae

Several members of this family are found associated with dried, stored foodstuffs. The best known species is the cosmopolitan saw-toothed grain beetle, *Oryzaephilus surinamensis* (L.), which attacks a wide range of grain and stored food products.

Family: Coccinellidae

Most of the Coccinellids are beneficial insects which prey upon aphids, scale insects and mites. Both adults and larvae are efficient predators and are important biological control agents. Some species have been introduced from one country to another with considerable success. In the early part of the century the introduction of the Australian *Rodiola cardinalis* Muls. into California was instrumental in controlling the cottony cushion scale *Icerya purchasi* on citrus. *Cryptolaemus montrouzieri* Muls. is another Australian species which has been used effectively for mealybug control in Hawaii and California.

The subfamily Epilachninae are, however, leaf-eating Coccinellids and some of these are serious plant pests. Both adults and larvae of the Mexican bean beetle, *Epilachna varivestis* Muls., feed on bean leaves in the United States and Central America. *Epilachna similis* (Thnb.), the maize ladybird beetle of Africa feeds on various wild and cultivated Gramineae; especially maize, pennisetum and sorghum. *Henosepilachna elaterii* (Rossi), the African melon ladybird, attacks cucurbits throughout Africa, the Mediterranean and the Near and Middle East.

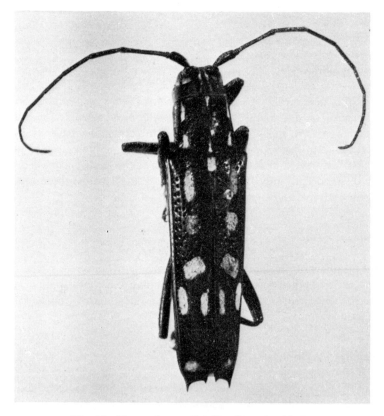

Fig. 19. *Glenea elegans* (family: Cerambycidae).

Family: Tenebrionidae

This large family of beetles contains few plant pests. *Gonocephalum simplex* (F.), the dusty brown bettle, is polyphagous. Larvae feed on plant roots in organically rich soil, and the adults attack both roots and stems of numerous vegetables and wild plants in the Sudan and other parts of tropical Africa.

There are several widely distributed but relatively minor pests of stored cereal products, e.g. *Gnathocerus cornutus* F., *Latheticus oryzae* Waterh., *Palorus ratzeburgii* Wissm., *Alphitobius diaperinus* Panz., *Tribolium castaneum* Herbst., *T. confusum* Duval, *Tenebrio molitor* L. and *Tenebrio confusus* F.

Superfamily: Chrysomeloidea

Family: Cerambycidae

The phytophagous larvae of this family usually bore in the wood of dead or dying trees but some feed in the stems or roots of herbaceous plants. *Hylotrupes bajalus* (L.) is an important wood borer of structural timbers (especially *Pinus* spp.). This species originated in Algeria but is now widely distributed. There are a number of borers of coffee, tea, cocoa, rubber and mango which are listed in Table 15 (see pp. 98–99). Some of these girdle the trunk or branch or weaken it sufficiently to cause it to die or break off. Among the borers of herbaceous plants, *Apomecyna binubila* Pasc., the melon stem borer, attacks cucurbits throughout Africa.

Family: Bruchidae

The larvae of seed "weevils" live mainly in the seeds of Leguminosae and Palmaceae. Some, such as the widely distributed *Acanthoscelides obtectus* (Say) attack beans and peas in the field as well as in store. Other are pests of growing peas. *Pachymerus longus* Pic. and *P. fuscus* Goeze larvae feed on unshelled groundnuts in Africa. *P. lacerdae* Chevr. attacks the nuts of oil palms.

Family: Chrysomelidae

Subfamily Cassidinae. Tortoise beetles of the genus *Aspidomorpha* feed on the leaves of sweet potato in the West Indies, New Guinea, Africa and the Pacific, both as larvae and adults. Another sweet potato pest is *Typophorus viridicyaneus* (Crotch) of the subfamily Eumolpinae. In this case the larvae feed on the underground tubers in Central America and the Souther United States. Some other Eumolopine root feeders are *Colaspis* spp. on cotton,

Table 15. Some Cerambycids boring in tropical economic plants

Species	Common name	Plant affected	Distribution
Anthores leuconotus Pasc.	White coffee borer	Coffee	Africa
Bixadus sierricola White		Coffee	Africa
Chreostes obesus Westw.		Coffee	Angola
Dirpha princeps Jord.		Coffee	Africa
Dirpha usambica Kolbe		Coffee	East Africa
Xylotrechus quadripes Chevr.		Coffee	India, Ceylon, Vietnam, Indonesia, Philippines
Xylotrechus javanicus Cast.		Coffee	Indonesia
Dihammus rusticola F.		Coffee	Sumatra

Species	Host	Location
Glenea novemguttata Guer.	Cocoa	Indonesia (cocoa sapwood)
Glenea aluensis Gah., *G.lefeburei* Guer.	Cocoa	New Guinea (cocoa sapwood)
Mallodon downesi F.	Cocoa	Africa
Monochamus ruspator F.	Cocoa	Africa
Steirastoma breve Guby	Cocoa	South America, West Indies
Tragocephala nobilis F.	Cocoa	Africa
Batocera rufumaculata DeG.	Rubber, mango	East Africa, India, Ceylon
Stenodontes downesi Hope	Rubber	Africa
Aolesthes induta Newm.	Tea	Taiwan, New Guinea

banana and Eucalyptus in South America. *Colaspis hypochlora* Lef., a pest of bananas in Central and South America, scars the fruit through adult feeding, thus reducing its market value. *Rhyparida morosa* Jac. feeds on grasses and attacks sugar cane in Queensland.

Flea beetles (subfamily: Halticinae) are very small beetles which are adult leaf feeders on a wide range of host plants. Eggs are usually laid in the soil and the larvae feed on roots and pupate in the soil. The adults, which can

Fig. 20. *Brontispa longissima* Gestro (family: Chrysomelidae, subfamily: Hispinae).

jump, often occur in large populations and produce characteristic small feeding holes over the surface of the leaf. The cabbage flea beetle, *Phyllotreta cheiranthi* Weise, attacks cultivated and wild Crucifers in Africa and Ceylon and may be a serious pest in the winter. *Podagrica puncticollis* Weise and *P. pallida* (Jac.) are pests of cotton in the Sudan and neighbouring countries. They attack the leaves of cotton and other Malvaceae as well as a variety of other hosts. Cotton seedlings may be completely destroyed by these insects in dry weather.

The subfamily Galerucinae also includes a number of injurious species. The red melon beetle, *Aulacophora africana* Weise, which occurs around the Mediterranean, Africa, the Near and Middle East and the USSR, is a particular pest of cucurbits but also attacks a number of legumes as well as cotton and tobacco. The larvae attack roots and subterranean parts of the plants while the adults feed vigorously on leaves and flowers. *Diabrotica* spp. attack cucurbits in the New World, while *D. vitta* F. is a vector of bacterial wilt and cucumber mosaic. *Buphonella murina* Gerst. is a pest of maize in East Africa, the larvae feeding in the lower part of the stem while the adults feed on the leaves. *Exora* spp. attack Sun hemp in Rhodesia. In West Africa beans and cowpeas are attacked by the adults of *Ootheca mutabilis* (Sahlb.).

Fig. 21. Coconut palm frond damaged by *B. longissima*.

The subfamily Chrysomelinae includes the Colorado beetle *Leptinotarsa decemlineata* (Say) which attacks potatoes and other Solanaceae in Europe, North and Central America. Several species of *Paropsis* feed on the leaves of *Eucalyptus* (both as larvae and adults) and these sometimes defoliate cultivated trees.

The subfamily Hispinae includes a number of important leaf miners or leaf feeders on palms. These may be divided into two groups: the leaf surface feeders and the leaf miners. The leaf surface feeders such as *Brontispa longissima* Gestro feed on the inner surfaces of unopened palm leaves and complete their life cycle before the leaf unrolls, resulting in a scorched and damaged appearance (Figs 20 and 21). Only young palms are attacked in New Guinea,

Fig. 22. *Promecotheca papuana* Csiki (family: Chrysomelidae, subfamily: Hispinae).

the Pacific and Indonesia. Similar damage is inflicted on coconuts in Malaysia by *Plesispa reichei* Chap.

The leaf mining habit is more widespread and is typified by *Promecotheca* where both larvae and adults feed inside the tissues of the unopened coconut leaves (Fig. 22). The adults chew parallel lines to the midrib, giving a striped effect. Some of the Hispid leaf miners of rice are of great economic importance, e.g. *Dicladispa armigera* (Ol.) which may cause losses of up to 50% of the rice crop in East Pakistan. It is particularly damaging to seedlings. Some Hispids of economic importance are listed in Table 16.

Table 16. Hispinae attacking some major tropical crops

Species	Leaf miner	Coco- nut	Oil palm	Rice	Sugar cane	Distribution
Brontispa longissima Gestro	–	*				Indonesia, New Guinea, Pacific
Plesispa reichi Chap.	–	*				Malaysia
Promecotheca cumingi Baly	*	*	*			Malaysia,Philippines
P.papuana Csiki	*	*	*			New Guinea
Coelaenomenodera elaeidis Mlk.	*	(*)	*			West Africa
Dicladispa armigera (Ol.)	*			*	*	India, Ceylon, South East Asia, New Guinea
Hispa stygia (Chap.)	*			*		India
Asamangulia wakkeri (Zehnt.)	*			*	*	Indonesia
Rhadinosa parvula (Motsch.)	*			*		Indonesia
R. lebongensis Maulik	*			*	*	India, China
Trichispa sericea (Guer.)	*			*		Africa, Madagascar
Leptispa spp.	*			*		India, Ceylon

Superfamily: Curculionoidea

Family: Anthribidae

This large family is well developed in the Indo-Malayan region. Most of the larvae are found in dead wood or fungi but *Araecerus fasciculatus* DeG. is a stored products pest of cocoa, coffee beans and spices. It is cosmopolitan in distribution.

Family: Apionidae

This family includes the important sweet potato weevil, *Cylas formicarius* (F.), of India and Southeast Asia which also occurs wherever sweet potatoes are grown. Another species, *C.puncticollis* Boh., is confined to Africa. The handsome red and black adult Cylas feed on the stems and large leaf veins of sweet potato and other fleshy Convolvulaceae. Most of the damage is, however,

Fig. 23. *Rhynchoporus bilineatus* (Montr.) (family: Curculionidae).

done by the larvae which bore into the tubers both in the field and in storage. Heavily infested tubers are bitter in flavour and are rejected by humans and stock as food.

The genus *Apion* includes several stem boring species attacking cotton and jute. The larvae are the damaging stage of the life history. Other species develop in plant roots or seeds.

Family: Curculionidae (Weevils)

This is the largest family in the animal kingdom and it contains many important tropical pests of crops, which attack all parts of the plants. Among the root feeders is the banana weevil *Cosmopolites sordidus* (Germ.) which bores into the rhizome of bananas, hemp and yams. This is one of the most serious pests of banana. *Euscepes batatae* Waterh. is a serious pest of sweet potato tubers in the West Indies and Central and South America.

Fig. 24. *Pantorhytes plutus* Oberth. (family: Curculionidae).

Among the most serious of the larval stem borers are the palm weevils *Rhynchophorus* spp. (Figs 23 and 25). Following injury by other insects (such as *Oryctes*) or human damage, the adult *Rhychophorus* lays its eggs within the palm stem. The developing larvae feed in the stem and destroy it causing serious economic losses. Though more limited in distribution, the cocoa stem borers (*Pantorhytes* spp.) (Fig. 24) may cause substantial losses of cocoa production in New Guinea. Larvae feeding in the jorquette region sometimes weaken the trunk sufficiently to cause the tree to split or to interfere with the flow of nutrients to and from the branches.

Fig. 25. Black palm weevil damage to coconut trunk.

Among the weevils attacking fruit and seeds the best known species is the Cotton Boll Weevil, *Anthonomus grandis* Boh., which is responsible for substantial losses in cotton production in the United States, Central America and the West Indies. The adults puncture the squares and bolls while feeding and lay their eggs inside them. The larvae feed on the developing flowers and destroy them or greatly reduce the production of seeds and cotton.

The grain weevil, *Sitophilus granarius* L., and the rice weevil, *Sitophilus oryzae* L., which are seed feeders, are among the most destructive pests of stored grain. The adults lay their eggs singly in chewed depressions on the

Fig. 26. Larva of black palm weevil, *Rhynchphorus*.

grain. The larvae feed inside the grains and hollow them out. The two species are similar in life history and often occur together.

The bark beetles and ambrosia beetles of the subfamilies Scolytinae and Platypodinae are small cylindrical insects. Many of them bore in the bark and sapwood, e.g. *Hylurdrectonus araucariae* Schedl, a major pest of hoop pine (Araucaria) in Papua New Guinea. Others bore in freshly cut timber, e.g. *Xyleborus perforans* Woll. and *Platypus jansoni* Chap. A few bore in fruits or seeds and among these is a serious pest of coffee, the coffee berry borer *Hypothenemus hampei* (Ferr.) which attacks different species and varieties of coffee as well as various members of the Rubiaceae, Malvaceae and Leguminosae. The adult female bores through the apex of green or ripening berries and then gnaws mines in the bean where it lays eggs. The larvae continue feeding within the bean and render it economically useless. This species is believed to be a native of Africa but it is widely distributed in Ceylon, South East Asia, West Irian, the Pacific and South America. Some other species of economic importance are listed in Table 17.

Table 17. Some weevils attacking tropical crops

Species	Common name	Host	Distribution
1. RICE			
Lissorhoptrus oryzophilus Kusch.	American water weevil	Serious larval damage to roots, some adult damage to leaves	Southern USA, Mexico. (Other spp. cause similar damage in Central and South America)
Helodytes foveolatus (Duval).			South America
Echinocnemus oryzae Mshl.		Aquatic larvae attack root hairs in clay soils	India
Tanymecus indicus Faust.		Adults attack growing point of young seedlings. Some root damage by larvae and adults	India
Hydronomidius molitor Faust.		Similar to above	Gujarat
2. COFFEE			
Cryptorhynchus sp.		Larvae tunnelling in trunk and ring-barking tree	New Guinea
Meroleptus cinctor Mshl.		As above	New Guinea
Apirocalus cornutus Pasc.		Adult feeding on growing points, of coffee, tea, avocado and others	New Guinea

Aulacophrys facialis Mshl.		Adult leaf feeder Polyphagous	New Guinea
Oribius spp.		Adults feed on leaves and berries	New Guinea
Diaprepes famelicus (Oliv.)		Roots attacked by larvae Also on sugar cane	West Indies
Lachnopus coffeae Mshl.		Adults feed on shoots, flowers and fruits	Central America and West Indies
Hypothenemus hampei (Ferr.)		Coffee berry borer	Pantropical
3. COTTON			
Alcidodes brevirostris (Boh.)	Cotton stem girdling	Adult feeds on stems	Africa
Apion soleatum Wagn.	Cotton stem weevil	Larvae boring at collar and nodes	Africa
Apion varium Wagn.		As above	Africa
Anthonomus grandis Boh.	Cotton boll weevil	Major pest, larvae and adults attacking flowers and fruit	North and Central America, West Indies
4. TEA			
Dicastigus mlanjensis Mshl.		Adults feed on leaves	East Africa
Hypomeces squamosus Herbst.		As above	Indonesia

contd.

Table 17. (*contd.*)

Species	Common name	Host	Distribution
5. SUGAR CANE			
Rhabdoscelus obscurus (Boisd.)	New Guinea sugar cane weevil	Stem borer, also attacks banana, sago, coconut oil palm, paw paw	New Guinea, Pacific, Celebes, Taiwan
Metamasius hemipterus (L.)	Rotten cane stalk borer,	Cane stalk also banana, and coconut	Central and South America, West Indies, West Africa
Rhynchophorus palmarum (L.)		Stem borer, also palms (mainly)	Central and South America, West Indies
Anacentrinus spp.		Larvae attack shoots and roots close to surface of ground	USA
Cosmopolites sordidus (Germ.)		Stem borer but mainly on bananas, etc.	South East Asia, Queensland, Philippines
Diaprepes abbreviatus L. and *D. famelicus* (Oliv.)		Larvae bore sugar cane roots and stalks. Adults feed on leaves of beans, citrus, etc.	Central and South America, West Indies
6. PALM WEEVILS			
Rhynchophorus palmarum (L.)	South American palm weevil	Coconut and other palms and sugar cane	Central and South America, West Indies

Rhynchophorus ferrugineus (Oliv.)	Red palm weevil	Coconut, date, oil, sago and other palms	India, South East Asia, New Guinea, Solomon Islands
Rhynchophorus phoenicis (F.)	African palm weevil	Palms	Africa
Rhynchophorus bilineatus (Montr.)		Coconut	New Guinea
Rhynchoporus vulneratus Panz.		Coconut	Malaysia
Metamasius hemipterus (L.)	Rotten cane stalk borer	Cane stalk also bananas and coconut	Central and South America, West Indies, West Africa
Diocalandra taitiense (Guer.)	Tahiti coconut weevil	Coconut	Madagascar, New Guinea, Pacific
Diocalandra frumenti (F.)	Four spotted coconut weevil	Coconut and other palms	Madagascar, Ceylon, South East Asia, New Guinea, Pacific
Amerrhinus pantherinus Oliv.	Palm leaf stalk borer	Palms	Brazil
7. VEGETABLE WEEVILS			
Cylas formicarius (F.)	Sweet potato weevil	Sweet potatoes	Pantropical
Cylas puncticollis Boh.		Sweet potatoes	Africa
Cylas brunneus (F.)		Sweet potatoes	West Africa
Euscepes batatae Waterh.		Sweet potato tubers	West Indies, South America
Alcidodes orientalis Mshl.		Larvae in sweet potatoes	East Africa
Baris traegardhi Aur.	Melon weevil	Melon. Larvae in fruits of cucurbits	North East Africa

contd.

Table 17 (*contd.*)

Species	Common name	Host	Distribution
Listroderes costirostris Schon.	Vegetable weevil	Vegetables	North and South America, South Africa, Australasia, Pacific
Graphognathus leucoloma (Boh.)	White fringed beetle	Many crops, vegetables and weeds	North and South America, Australasia, mainly temperate
Pantomorus cervinus (Boh.)	Fuller's rose weevil	Polyphagous	North and South America, Mediterranean North and South Africa, Australasia, mainly temperate
WEEVILS ATTACKING OTHER PLANTS			
Sternochetus frigidus F.	Mango weevil	Mango	New Guinea, South East Asia
Sternochetus mangiferae (F.)	Mango stone weevil	Mango	Africa, Madagascar, Pakistan, India, Ceylon, Malaysia, Philippines, Queensland, Pacific
Rhynchaenus mangiferae Mshl.	Mango leaf miner	Mango	India
Deporaus marginatus Pas.	Mango leaf weevil	Mango	India, Ceylon
Cosmopolites sordidus (Germ.)	Banana weevil	One of the most serious banana pests, also attacks hemp, plantain and yam	South East Asia, Philippines, Queensland

Species	Common name	Host	Distribution
Apirocalus cornutus Pas.		Attacks banana leaves and flowers, polyphagus	New Guinea
Odoiporus longicollis (Oliv.)	Banana stem borer weevil	Bananas	South East Asia, Cylon to Taiwan
Sphenophorus striatus F.	Cocoa stem borer	Cocoa	Fiji, San Thome
Pantorhytes spp.	Cocoa stem borer	Cocoa	New Guinea, Solomon Is.
Coelosternus spp.	Cassava stem borer	Cassava	West Indies, South America
Cratosomus punctulatus Gyll.		Citrus borer and leaf feeder	West Indies, Central America
Metamasius ritchiei Bs.	Pineapple fruit stalk borer	Pineapple	West Indies
Metamasius hemipterus (L.)		Banana, coconut, sugar cane borer	Central and South America, West Indies, West Africa
Apion corchori Mshl.	Jute stem borer	Jute	India
Scyphophorus interstitialis Gyll.	Sisal leaf weevil		Americas, East Africa
Vanapa oberthuri Pouill.	Hoop pine borer		New Guinea
Omophorus stomachosus Boh.	Fig fruit weevil		South Africa, West Africa
Balanogastris kolae Desbr.		Kola	West Africa
Gonipterus scutellatus Gyll.	Eucalyptus leaf feeder	Eucalyptus	Australia, Africa, South America

contd.

Table 17 (*contd.*)

Species	Common name	Host	Distribution
Hypera postica (Gyll.)	Alfalfa weevil		USA, Europe North Africa, Middle East, to Indian and Pakistan
9. WEEVILS INFESTING STORED PRODUCTS			
Sitophilus oryzae L.	Rice weevil	Wide variety of cereals	Cosmopolitan
Sitophilus granarius L.	Granary weevil	Wide variety of cereals	Cosmopolitan
Caulophilus latinasus Say.	Broad-nosed granary weevil		USA, Central America, West Indies
Araecerus fasciculatus DeG.		Coffee and cocoa beans and spices	Cosmopolitan
10. PLATYPODID AND SCOLYTID WEEVILS ON TROPICAL CROPS			
Xylosandrus morigerus (Bldf.)	Brown coffee borer	Orchids, coffee, mahogany	New Guinea, Indonesia, Pacific, Malaysia, Ceylon, Madagascar, South America
Xylosandrus compactus (Eich.)		Coffee, cocoa, avocado	Malaysia, Indonesia, Vietnam, India, Ceylon
Xyleborus ferrugineus (F.)		Rubber, cocoa, coconut	Americas, Africa, Malaysia

Species	Common name	Host	Distribution
X. affinis Eich.	Rubber shot hole borer	Rubber	Africa, Indonesia, Americas
X. fornicatus Eich.	Tea shot hole borer	Tea, coffee	Africa, Indonesia, Ceylon
X. discolor Bldf.		Coffee, cocoa, cinnamon	India, Ceylon, Vietnam, Taiwan, Indonesia
X. habercorni Egg.		Coffee	South East Asia
X. bicornis Egg.		Coffee	Sumatra
X. brasiliensis Egg.		Coffee	Brazil
Hypothenemus hampei (Ferr.)	Coffee berry borer	Coffee (fruits)	Africa, Ceylon, South East Asia, West Irian, South America
Hypothenemus areccae Horn		Coffee (fruits) betel nuts	India, South East Asia, Pacific, South America, Africa
Hylurdrectonus araucariae Schedl.		Hoop pine	New Guinea
Xyleborus perforans Woll.		Freshly cut timber	New Guinea, Pacific, South East Asia, Madagascar, North and South America
Platypus jansoni Chap.		Freshly cut timber	New Guinea, Indonesia

Many weevils are indiscriminate root feeders in the larval stage while the adults feed on the leaves and stems of a wide range of host plants, e.g. *Apirocalus cornutus* Pasc. (Fig. 27) which is sometimes damaging to banana leaves and flowers but is also polyphagous in New Guinea. Some other polyphagous pests (especially of vegetables) are the white fringed beetle, *Graphognathus leucoloma*, (Boh.) and the vegetable weevil, *Listroderes costirostris* Schonh. but these are mainly temperate in distribution.

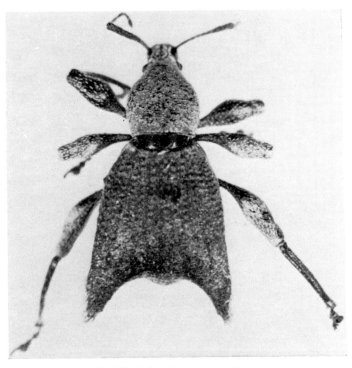

Fig. 27. *Apirocalus cornutus* Pasc.

Another widespread, polyphagous species which sometimes attacks Citrus is Fuller's rose weevil, *Pantomorus cervinus* (Boh.). This appears to be absent from India and South East Asia but is widely distributed elsewhere.

FURTHER READING

Crowson, R. A. (1955). *The natural classification of the families of Coleoptera*. Nathaniel Lloyd, London, pp. 186.

Hinton, H. E. (1945). *A monograph of the beetles associated with stored products.* British Museum (Nat. Hist.), London, pp. 443.

Jeannel, P. and Paulian, R. (1949). Ordre des Coleopteres in Grassé P. (Ed.). *Traité de Zoologie.* Vol. 9. Masson et Cie, Paris, pp. 771–1077.

Schenkling, S. (Ed.) (1910–1940). *Coleopterorum Catalogus* Vols. 1–31 with supplements. W. Junk, The Hague.

Chapter 11   *Hymenoptera*

This enormous group with 120,000 species is the second largest order in the animal kingdom. It includes more than 150 families which are usually divided into two suborders. The more primitive suborder Symphyta includes a number of phytophagous species but few of major economic significance. The other suborder, *Apocrita*, includes a great diversity of forms but few of them are plant feeders. However a great number of insects of benefit to man occur in this group, including the main plant pollinators and a number of predatory and parasitic wasps which play important roles in the regulation of insect numbers.

Suborder: Symphyta

In the adult the thorax is broadly joined to the abdomen. The larvae often resemble caterpillars but have a single pair of ocelli while larval Lepidoptera typically have six pairs of ocelli.

Superfamily: Siricoidea

Family: Siricidae

The wood wasps or horntails are large, brightly coloured insects. The female has a powerful, piercing ovipositor with which eggs are laid in the young wood of trees. The widely distributed *Sirex noctilio* F. mainly attacks injured or weakened pine trees. The galleries formed in the wood by the larvae render the timber useless. This is mainly a pest of temperate regions.

Superfamily: Tenthredinoidae

The family Diprionidae includes a number of pests of coniferous trees. In the family Pergidae the larvae of several genera of sawflies feed on Eucalyptus leaves and sometimes cause complete defoliation. The larvae of *Caliroa* and some other genera of the family Tenthredinidae skeletonize the leaves of fruit trees and ornamentals.

Suborder: Apocrita

This includes the vast majority of Hymenoptera including the ants, wasps and bees. In the adult the basal part of the abdomen is constricted or stalk-like

The larvae, many of which are internal or external parasites, are legless and have a distinct head capsule. The 15 superfamilies can be broadly grouped into those which are parasitic (e.g. Ichneumonoidea, Chalcidoidea), predatory (some Pompiloidea, Vespoidea and Sphecoidea) and those which are secondarily parasitic or have changed to plant food (some Vespoidea, Sphecoidea and Formicoidea). Social behaviour reaches its highest development in this order and this has probably arisen independently at least three times in the course of evolution.

The superfamily Ichneumonoidea, which includes the families Ichneumonidae and Braconidae parasitize a very wide variety of insects though Lepidoptera are particularly extensively attacked. Coleoptera and Diptera are less fequently attacked while one subfamily of Braconids are important parasites of aphids.

The superfamily Proctotrupoidea is a large group of small parasites, including a number of egg parasites as well as some hyperparasites.

The superfamily Cynipoidea includes a number of parasites of Diptera pupae and secondary parasites of aphids. The family Cynipidae includes a number of species which form galls on plants.

The very large superfamily Chalcidoidea mostly comprises tiny parasitic or hyperparasitic insects but there are some plant feeding forms of minor economic interest including a number of gall formers. One family, the Agaonidae, are of interest for their role in the pollination of figs. These insects live inside the fruits and the wingless males rarely emerge from them. Those hyperparasites which attack useful parasites must be regarded as injurious from the economic point of view.

The small superfamilies Chrysidoidea and Bethyloidea comprise mainly ectoparasitic species. The superfamilies Pompiloidea, Scolioidea and Sphecoidea are all digging wasps and are predatory or parasitic. The Pompiloids feed on spiders while many of the stoutly built Scolioids parasitize scarab larvae in the soil. Other Scolioids parasitize ground-dwelling sphecoid and vespoid wasps and bees. The Sphecoidea are solitary, predatory hunting wasps. The females feed their prey to larvae which often live in nests excavated in the soil or trees or in mud cells attached to buildings or trees.

The superfamily Vespoidea includes some solitary hunting wasps but in the family Vespidae are found the social predatory wasps which construct nests underground or in trees. The nests are constructed of paper made from chewed wood or bark and the combs which are arranged in layers may include many thousand larvae in a single nest. Since these are fed on caterpillars and small insects, the predatory activities of the adults may be very substantial. Many of the wasps can inflict a painful sting and multiple stings of some of the larger hornets can be fatal to man.

The superfamily Sphecoidea includes both solitary and colonial wasps. The adults feed on sugary plant or insects secretions but the females are active predators on insects which are fed to their larvae. The members of the large family Sphecidae are important predators which construct mud cells in the ground or on rocks, trees or buildings. These mud daubers are abundant in the tropics and while they have an important predatory role they are sometimes a nuisance in houses by causing damage to books and fabrics while constructing their nests.

The superfamily Apoidea, which includes the bees, are mostly solitary forms but some well-known social ones occur as well. These resemble the sphecoid wasps but for larval protein food they use pollen instead of animal prey. Their main source of carbohydrate is nectar from flowers. These biological developments have rendered the bees among the most important insects for agriculture because of their role as plant pollinators.

Members of the family Megachilidae use foreign materials in the construction of their solitary burrows. In the genus *Megachile* the leaf cutter bees cut discs of leaf tissue for this purpose and are sometimes minor agricultural pests. Some small Halictidae (sweat bees) are a nuisance when they land on the skin in large numbers to feed on perspiration. However, many of the members of this family are useful pollinators as are many Andrenidae and Anthophoridae. The large, robust carpenter bees (*Xylocopa*), which are included in the Anthophorids, produce large nests in dried wood and can cause considerable damage in fence posts and structural timber in the tropics.

The family Apidae, which includes the social honeybees and bumblebees is of great economic importance. The long-tongued bumblebees are important pollinators of clovers, alfalfa and other legumes but they are not widespread in the tropics except at higher altitudes. The honeybee has long been sought out and cultivated by man. The most primitive hunting and gathering people seek out honeycombs as a delicacy. Three of the four species of honeybees (*Apis dorsata* F., *A. florea* F. and *A. indica* F.) are natives of India, Ceylon and South East Asia while the domesticated western honeybee (*Apis mellifera* L.) is found in the wild state in Europe. Only *A. mellifera* and *A. indica* (which are tree-nesting species) have been domesticated and persuaded to nest in hives. A number of subspecies and races of *A. mellifera* are recognized. These polymorphic insects exhibit advanced social behaviour. The waxen nest (comb) is constructed by the workers who are infertile females. Honey which is stored and fed to other members of the colony is manufactured from nectar and insect honeydew in the honey stomach of the worker. Pollen, water and a salivary secretion of the workers (royal jelly or brood food), provides all the other nutrients required by the colony. The sale of honey and beeswax is a large industry but more more important is the role of bees in pollination of trees and crop plants. The indiscriminate use of insecticides can wipe out bee populations and this is a factor to be considered when planning spray programmes. Bees are also attacked by parasitic mites which can destroy the colony. *Acarapis woodi*, an internal parasite of the tracheal system, is particularly damaging.

Despite the great importance of the Apoidea as plant pollinators, it must be remembered that many other insects also fulfil this function, especially amongst the Lepidoptera and Diptera, e.g. the main pollinator of cocoa in some areas is the midge *Culicoides*.

Superfamily: Formicoidea

Family: Formicidae

The ants are social insects of ubiquitous occurrence. Considering their diversity and abundance in the tropics it is hardly surprising that they are of great importance in ecosystems of interest to the economic entomologist. Nests are commonly found underground or in cavities in trees, buildings or rocks. Some species, e.g. *Oecophylla* (weaver ants) build nests of leaves joined by larval silk in trees.

Several different forms occur within nests: the winged male and sexual female (queen) and the wingless infertile female worker castes. Many ants feed on both plant and animal material but some are mainly carnivorous (e.g. in the subfamilies Ponerinae and Dorylineae), while others are mainly herbivorous and feed on seeds, fungi, and plants. The predaceous species play an important role in biological control. If however they attack the parasites or predators of another pest, this role can be harmful.

The driver ants of the subfamily Dorylinae are mainly carnivorous on other insects but may attack mammals or birds encountered during their marches. These ants are characterized by their massive predatory raids against other insects and great nomadic movements of the colonies. They are able to form and readily abandon temporary nests and all colonies are able to change reversibly from high to low activity levels.

The tapinoma ants of the subfamily Dolichoderinae includes some annoying domestic species such as the Argentine ant, *Iridomyrmex humilis* Mayr. A more undesirable characteristic of this and other members of the family (as well as a number of Myrmicinae) is the tending of aphids and coccids for their honeydew. This discourages the natural enemies of the Homoptera and leads to an increase in their populations and indirectly to a spread of some plant virus diseases.

In the subfamily Myrmicinae the fire ant *Solenopsis geminata* (F.) is a serious pest throughout the tropics. This polyphagous ant damages citrus and other fruit trees by girdling the bark. It also destroys seedlings, removes seeds and tends aphids, scales and mealybugs. It is an aggressive and dominant species on sugar cane, coffee, cocoa, citrus and other fruit trees in the West Indies. The leaf cutting parasol ants of the genus *Atta* cut portions from leaves and carry them to their underground nests where they are chewed to form a substrate for the fungi which provide the main food for the colony. They can be quite damaging in plant nurseries. *Messor barbarus* L., the harvester ant of the Mediterranean and Africa, collects grass and cereal seeds and destroys the vegetation near its nest. It is a severe pest in light, sandy soils.

Fig. 28. The ant *Oecophylla smaragdina* attacking adult
Pantorhytes on cocoa.

The subfamily Formicinae is well developed in tropical and subtropical regions. Many of them are soil dwellers and live in very large colonies. Feeding on nectar or honeydew is widespread and some are specialized for the storage of this sugary material (honeypot ants). The red and green tree dwelling ants of the genus *Oecophylla* are aggressive biters which tend aphids and coccids and have a significant effect on the insect fauna of infested trees: sometimes beneficial and sometimes not (Fig. 28).

Leston (1970) has drawn attention to the dominant role of ants in the entomological problems of tropical tree crops. In West African cocoa there was a mosaic of species including three mutually exclusive elements. These ants were important in controlling the predators of capsids and the abundance of mealybugs. Since the composition of the ant fauna is influenced by cultural practices (such as shade) as well as by insecticidal spray programmes, it is apparent that this important part of the ecosystem needs careful consideration when control studies are made. It has been shown that ants are an important factor in the transmission of swollen shoot virus disease of cocoa and that appropriate spray programmes directed against the right species of ant may be more effective than sprays applied to the mealybug vector.

This stresses the need for more detailed ecological studies of tropical crops and associated plants in order to design effective control strategies.

FURTHER READING

Berland, L. and Bernard F. (1951). Ordre des Hymenopteres in Grassé P. (Ed.) *Traité de Zoologie*. Vol. 10. Masson et Cie, Paris, pp. 771–1276.
Clausen, C. P. (1940). *Entomophagous insects*, N.Y. pp. 688.
Doutt, R. L. (1951). The biology of parasitic Hymenoptera. *Ann. Rev. Ent.* **4**. 161–182.
Nixon, G. (1954). *The world of bees*. London, pp. 215.
Ribbands, C. R. (1953). *The behaviour and social life of bees*. London, pp. 352.
Sudd, J. H. (1967). *An introduction to the behaviour of ants*. London, pp. 200.

Chapter 12   Ecology of Pest Control

The main aim of the farmer should be the management of his agro-ecosystem in such a manner as to ensure continuing and economic crop yields without damage to the ecosystem. For this to be accomplished it is often necessary to protect plants against attacks of insect pests. Insects only become agricultural pests when their activities interfere with human welfare. Under natural conditions, whether in the jungle canopy or on the savanna, the mechanisms of evolution acting over a very long period of time have eliminated the less well-adapted organisms and given a condition of unstable equilibrium in the survivors which is easily upset. In setting out to control a pest the aim is often to reduce the numbers of the insect concerned, but control means rather more than killing unwanted insects. To approach the problem in a rational manner we must consider how and why some insects have come to be regarded as pests.

Generally speaking four situations can be recognized:

(1) the entry of a new insect species into a region;

(2) changes in the characteristics of a species which formerly did not compete with man directly;

(3) changes in human activities or habits which make us sensitive to the existence of species which formerly occurred in such small numbers as to be unimportant;

(4) increase in insect abundance following changes in the environment. The increase could result from an improved supply of a limiting resource or from the decrease of antagonistic interactions that formerly prevented the species from fully exploiting its environment, or from both sets of factors.

All phytophagous insects are potential pests but they may be classified into four groups from the practical point of view:

(1) serious perennial pests;

(2) intermittent but sometimes serious pests;

(3) insects that are rarely ever pests;

(4) non-residents of the ecosystem that periodically enter the environment for short times but then are serious though irregular or rare pests.

Serious perennial pests are commonly encountered in orchards and plantations and may require continuing sprays for control.

The intermittent pests of group (2) vary in importance. Although they are potentially important they do not always realise their potential, e.g. aphids, scale insects, mites and leaf rollers. The proper planning of control measures is very important here.

The insects of group (3) are seldom pests because they are not abundant or because they can be tolerated at high densities. Unsuitable spray treatments can sometimes convert these insects into serious pests.

The insects which are non-residents of the ecosystem but enter it occasionally to feed (such as some capsid bugs) are often difficult to control.

Any student of insect ecology is constantly reminded that reproductive potential greatly exceeds survival in any species. The entry of a species into a new region without the natural enemies which normally restrict its numbers often results in an upsurge of population. We cannot consider pests apart from their environment. The pest populations are part of ecosystems which include complexes of plant and animal populations as well as the physical environment in which the live. In other words the environment includes not only the non-living surroundings but also other living organisms as well as individuals of the same species which share the surroundings. Population numbers are determined both by the intrinsic qualities of an organism provided by its heredity and by the qualities of the environment. The ecosystem is therefore complex in structure and Clark *et al.* (1967) recommend that we should study the "life system" of the population which they define as that part of the ecosystem which determines the existance, abundance and evolution of a particular population. Translated into practical terms this means that the economic entomologist has to extend his investigations far beyond the pest species that he is investigating if he is to appreciate the reasons for fluctuation in its numbers.

One means of upsetting the equilibrium of insect populations is to grow large numbers of plants of the same kind together (monoculture); this is of course a common agricultural practice. It has the effect of providing large quantities of food for herbivores and incidentally makes conditions less congenial for natural enemies by providing fewer alternative host plants for them to live on when they are not feeding on the pest species.

Many of the food crops cultivated today have been selected by man for many generations for improved yield, flavour or nutritional value but the same attention has not been given to selection for resistance to pests and diseases. Thus many of our crops are highly susceptible to attack, more so as a rule than are their wild type ancestors.

For a rational approach to pest control a thorough understanding of the ecology of both the host plant and the insect pest is necessary. Unfortunately this information is seldom available in the tropics. Sometimes the ecological situation is extremely complex as in the case of degeneration of cocoa in West Africa which involved virus diseases, capsids, mealybug vectors and ant attendants, not to mention host varieties and cultural practices.

The rate of population growth is determined by the difference between birth rate and death rate of the pest. When faced with a pest population the economic entomologist should aim at decreasing the birth rate and increasing the death rate of the species concerned, however, the means by which this is most effectively achieved may be indirect or subtle. Some of the factors bringing about increase in insect populations are abundant food, the absence of natural enemies and disease, favourable weather conditions and good synchronization with the life cycle of the host. On the other hand insect populations are decreased by insufficient or unsuitable food, by the presence of natural enemies and disease and by imperfect synchronization with the life cycle of the host. Many of these environmental factors operate on both the birth rate and the death rate. What can the economic entomologist do in practice?

The first problem is to recognize the existence of a pest situation and to identify and study the characteristics of the species concerned. This is not necessarily a straightforward procedure, e.g. in the case of long term progressive degeneration of crops associated with insect borne diseases or where small, fugitive populations of insects with highly toxic saliva are involved. Sound systematic identification is important because it frequently makes accessible the experience of other people through the literature, e.g. the *Review of Applied Entomology* and other abstracting journals. Many tropical entomologists working in relative isolation waste a great deal of time repeating other people's unsuccessful experiments. Intensive observation of the pest species in the field is indispensable but it is much more valuable if the results can be weighed against the findings (or omissions) of other workers.

Having recognized the existence of a pest situation and identified the insects concerned, the next step is to gather together information on the characteristics and life history of the species and its interactions with the environment (in the broad sense). This provides a basis for the practical assessment and analysis of the situation which can lead to a choice of the most promising strategy for control.

It has been suggested (Clarke *et al.* 1967) that there are four basic strategies in pest control:

(1) evasion of the consequences of pest activity;

(2) the elimination of those characteristics of the target which make it susceptible to attack;

(3) suppression of the characteristics and attributes of the pest that make it injurious; and

(4) reduction of pest numbers to a non-injurious level.

An example of (1) and (3) would be the growing of crops in areas where specific pests do not thrive. This method has been used for many years for growing seed potatoes in areas where potato virus spread is minimal due to the limited breeding and movement of vector species and the absence of local sources of virus infection.

Host plant resistance is an excellent example of (2) and this is considered in more detail in the next chapter. The reduction of numbers of pests to a non-injurious level (4) is the most common strategy through the use of insecticides and biological control.

A great deal has been written about the growth and fluctuation of size of insect populations. From the practical point of view it may be sufficient to point out that although insect numbers tend to fluctuate up and down, in the field they tend to move about an equilibrium position unless influenced by some external factor. This *general equilibrium position* is the average density of the population over a period of time in the absence of permanent environmental change. This usually lies below the *economic threshold* which is the density at which control measures should be taken to prevent an increasing pest population from reaching the *economic injury level*. This is the lowest population density causing sufficient economic damage to justify the cost of artificial control measures (see Fig. 29). The magnitude of these levels will depend upon the crop and pest concerned and the economic value of the crop. The choice of an appropriate economic injury level is of central importance for the design of a control programme and ideally requires the combined skills of the farmer, the entomologist and economist.

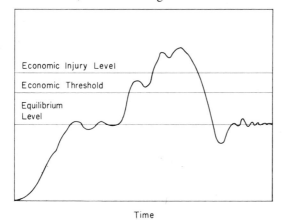

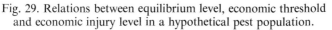

Time

Fig. 29. Relations between equilibrium level, economic threshold and economic injury level in a hypothetical pest population.

It is seldom feasible to eradicate pests completely. The economic entomologist usually aims to keep pest numbers within reasonable limits by creating such conditions within life systems that will reinforce the natural subtractive agencies. This has been termed *pest management*.

The reduction of pest numbers to a non-injurious level by the use of insecticides (Chapter 14) is the most common method of pest control (though not necessarily the most desirable). Populations may respond in one of three ways following the application of insecticides. If the mortality exceeds a critical limit they may die out immediately: this is unusual. Often the populations fluctuate in numbers in phase with the treatment—sometimes violently. The continued use of insecticides over a long period of time usually results in the need for more frequent applications to deal with these fluctuations. Alternatively the insect populations may reveal unexpected properties—such as the development of insecticide resistance. This has been a widespread phenomenon since World War II following the introduction of powerful, broad-spectrum insecticides, i.e. insecticides acting indiscriminately on a wide range of species both beneficial and harmful.

It has become increasingly obvious in recent years that these indiscriminate measures are undesirable because of their long term effects on parasites and predators which play an important role in the natural regulation of populations.

Although the agents of biological control virtually never exterminate an insect pest they will often reduce numbers to below the economic threshold for most of the time. Unfortunately parasites have their own parasites (hyperparasites) and these reduce their efficiency for biological control. Parasites and predators are of course susceptible to the action of many insecticides. However, since insects behave differently towards different insecticides it is often possible to select insecticides (and timing of applications) so that they will have a maximum effect on pest species and a minimum effect on its natural enemies. The search for more specific insecticides and the study of their persistence are important aspects of modern pesticide research. Resistant hosts plants tend to support small insect populations and the combination of host resistance with biological control enhances the efficiency of both. Combining these with appropriate cultural practices and the proper use of selective insecticides permits the development of *integrated control* programmes.

It is not possible to transplant the concepts and methods of temperate country pest control into the tropics. Greater numbers of pest species are encountered in the tropics and the rather uniform climatic conditions tend to favour insect population growth. At high temperatures it is difficult to persuade spray operators to wear adequate protective clothing and human toxicity hazards are higher. The main difference between zones however lies in differences between the agro-ecosystems. Particularly in small scale subsistence agriculture a mixture of crops are grown, usually in relatively small plots and for a limited length of time. Although the plant varieties are frequently low-yielding (and sometimes of indifferent quality) they usually have a high degree of genetic resistance to local pests and diseases (following many generations of natural selection). Furthermore this mixed cultural system frequently supports a wider range of insect parasites and predators than is found in monocultural systems. Thus these crops are less readily exploited by endemic pests.

When attempts are made to improve food production by the introduction of higher yielding varieties, the use of fertilizers and pesticides and other changes in cultural practice there is a radical disturbance of the agro-ecosystem which frequently leads to the appearance of new and severe pest problems.

This has been dramatically demonstrated in recent years by the "green revolution" which has greatly increased cereal yields in a number of developing countries. This increase has been brought about by the introduction of higher yielding plant varieties, a shift towards monoculture and widespread changes in agricultural practice including multiple cropping, irrigation and increased use of fertilizers and pesticides. These changes are ones which render the ecosystem more susceptible to large-scale pest and disease development. It is therefore important that entomologists and agriculturists in developing countries should be prepared for this and not automatically reach for the spray pump without giving careful consideration to the development of integrated systems of control which will be economical, effective and longlasting.

Our present capabilities of dealing with pests are often limited by lack of information. Even the quantitative evaluation of insect damage is difficult to obtain in a suitable objective form. In most cases we have too little knowledge of life systems to allow us to choose the most appropriate control strategies or to design integrated control programmes. More basic ecological knowledge is required for understanding the complex interrelationships between species encountered in the field. In many tropical countries this can only be obtained by intensive and detailed investigation on the spot. There is an urgent need for more research in these areas in the tropics.

FURTHER READING

Clarke, L. R., Geier, P. W., Hughes, R. D. and Morris, R. F. (1966). *The ecology of insect populations*. London, pp. 232.

Clausen, C. P. (1958). Biological control of insect pests. *Ann. Rev. Ent.* **3**. 291–310.

De Bach, P. (Ed.) (1964). *Biological control of insect pests and weeds*. Chapman and Hall, London, pp. 844.

Smith, R. F. (1972). The impact of the green revolution on plant protection in tropical and subtropical areas. *Bull. Ent. Soc. Amer.* **18** (1) 7–14.

Smith, R. F. and Van der Bosch, R. (1967). Integrated control in Kilgore, W. W. and Doutt, R. L. (Ed.) *Pest control*. Academic Press, New York, pp. 295–340.

Stern, V. M. (1973). Economic thresholds. *Ann. Rev. Ent.* **18**. 259–280.

Stern, V. M., Smith, R. F., Van der Bosch, R. and Hagen, K. S. (1957). The integrated control concept. *Hilgardia* **29**(2). 81–101.

Swezey, O. H. (1936). Biological control of the sugar cane leafhopper in Hawaii. *Bull. Exp. Sta. Hawaii Sugar Planters Assoc. (Ent. ser.) no.* 21, pp. 101.

Wood, R. K. S. (Ed.) (1960). Biological problems arising from the control of pests and diseases. Symposium, Institute of Biology. No. 9. London.

Chapter 13

Host Resistance and Susceptibility: Genetic Controls

Those plant parasitic insects which are confined to one species of host plant are said to be *monophagous*, those which are restricted to a small group of host species are said to be *oligophagous* while those which feed on a wide range of host plants from several plant families are termed *polyphagous*. The most common condition is oligophagy. This implies that at some stage in their life history (probably the adult) an insect has some means of host discrimination and appropriate patterns of behaviour which result in the rejection of non-host plants and/or the acceptance of host plants. This is of considerable importance in relation to host plant resistance to insects.

Within a population of plants of the same species some plants will support smaller populations of insects than others or will be less damaged by insect populations of the same size. These plants may be said to show some resistance to insect attack. Extending this to crop varieties within a species it is found that some varieties consistently exhibit a degree of resistance and furthermore this resistance can be genetically transferred to their offspring.

Thus *host resistance* does not imply immunity from insect attack but it does imply an increased probability of surviving an attack which would severely damage a susceptible variety. Often a number of genes are involved in host resistance. If these can be located by the plant breeder and transferred to agronomically suitable plant varieties by hybridization procedures we have a means of protecting crop plants from insect attack with a minimum recurrent expenditure. This technique has in fact been used with considerable success on a number of crop plants and its advantages are obvious. Resistant varieties are usually tested by exposing them to the appropriate pest species either in the glass house or in the field. This empirical method reflects our ignorance of the nature of resistance mechanisms in most cases. It has been suggested that three sets of factors are involved in host resistance:

(a) *non-preference* of the plants for oviposition, shelter or food because of the absence of certain qualities, perhaps chemical in nature;

(b) the resistant plant may have adverse effects on the biology of the insect. Painter termed this *antibiosis*;

(c) resistant plants may be *tolerant* and capable of surviving levels of infestation which would kill less robust plants.

These sets of factors are only a first approximation and are themselves complex. They are operationally useful concepts and help to sort out our ideas when investigating host resistance. They are not an explanation of resistance. It is obvious that the physiology and behaviour of host selection play an important role.

In 1920 it was observed that certain strains of American upland cotton were resistant to the cotton jassid *Empoasca*. These strains were not of good commercial quality but subsequent work by plant breeders resulted in the production of good commercial strains resistant to jassid attack. Resistance was found to be associated with (but not necessarily caused by) hairiness of the cotton stem. In the United States many years of selection and breeding have resulted in the production of several varieties of wheat resistant to the hessian fly. In New Zealand a variety of swede turnip has been developed which is resistant to aphids as well as to certain plant viruses and fungi.

Sometimes it is possible to breed for plant characteristics which interfere with insect feeding or egg-laying patterns. Thus work on resistance of corn against *Heliothis armigera* was directed towards producing plants with long, tight fitting husks and hard textured kernels over the tip of the ear.

Solid stem varieties of wheat were bred which were resistant to the wheat stem sawfly which normally feeds inside the stem. Sorghum varieties which were resistant to midge attack bore flowers which did not open for pollination —thus preventing midge oviposition. The cotton boll weevil *Anthonomus grandis* causes most damage late in the season when populations are large. Attention was given to the breeding of early and rapid fruiting cottons which set early when populations of boll weevil were small, thus avoiding attack and damage.

In searching for resistance it is sometimes necessary to seek out the wild ancestors or wild relatives of cultivated plants since they are most likely to show resistance to pests and diseases after many centuries of natural selection. Vavilov and his Russian colleagues first organized expeditions to collect wild relatives of crop plants and they made extensive studies of plants of the genus *Solanum* related to the domestic potato. This kind of investigation is invaluable for studying resistance to other types of plant disease. Hawaiian sugar planters have sent several expeditions to Papua New Guinea in search of sugar cane varieties which might be used for breeding purposes. Although sugar cane is not grown commercially in Papua, New Guinea, it is a centre of diversity and may well have been a centre of origin for sugar cane. Searches have also been made for the parasites of economic insects on sugar cane since it is likely that parasites will be found in old areas where insect populations are in equilibrium. Successful parasites can then be introduced into commercial plantings in other countries.

Although some cases of resistance involve one or a few genes which are inherited in a simple mendelian fashion, the situation is often more complicated and many genes are involved. Sometimes in searching for resistance the plant breeder accelerates the rate of mutation of plants by exposing them to radiation or mutagenic chemicals. Most of the resulting mutations are useless, but occasionally a useful one appears. Some cultivated plants are sterile or produce hybrids with difficulty. Grafting has been a useful technique in some such cases, e.g. rootstock grafting of grapevines for control of *Phylloxera* was a dramatic and effective control of a pest which threatened to wipe out the French wine industry. Similarly, some apple varieties are highly resistant to the attack of the woolly aphis *Eriosoma lanigerum* Hausm. and these can be used as root stocks for growing highly susceptible apple varieties.

The breeding of resistant varieties of commercial crops is a slow and tedious undertaking, often extending over many years. An important early step in the process is the selection of plants which show promising resistance in the field. The tropical entomologist should always be on the lookout for plants of this kind. It may mean only one or two plants in a whole crop but they should be carefully put aside and multiplied for future studies. This field selection may be invaluable for future work—and indeed farmers have carried out this kind of selection for centuries.

The *genetic control* of insects is directed at the pests themselves rather than at their hosts and with the aim of depressing their powers of multiplication and dispersal. Perhaps the most classical case was the use of sterile males for the control of screw worms of livestock in Curacao and the United States. Since these insects mate only once, it was reasoned that if the population could be flooded with sterile males there should be a dramatic decline of reproduction. Screwworms were bred in large numbers (many millions per week) and were sterilized by radiation while they were developing. The adults were released in large numbers and in several places the screw worm was completely eradicated. For this type of programme to be successful there must be minimal or no immigration and it must be possible to breed and release very large numbers of sterile individuals for several generations. These conditions may be difficult to satisfy and require considerable expenditure. The sterile males must greatly outnumber normal males so the technique is more efficient if the pest population can be reduced at the outset by the use of insecticides or other means.

Other attempts at genetic control have involved the introduction of deleterious genes into insect populations. These are still in the experimental stage but show some promise for eradication of some mosquitoes.

FURTHER READING

Anon. (1969). Insect pest management and control. *Principles of plant and animal pest control*, Vol. 3. pp. 508. Nat. Acad. Sci. Washington, plub. 1965.

Fraenkel, O. H. and Bennett, E. (Ed.) (1970). Genetic resources in plants. *IBP Handbook no.* 11, London, pp. 554.

Painter, R. H. (1951). *Insect resistance in crop plants.* University of Kansas Press, U.S.A., pp. 520.

Pathak, M. D. (1969). Stem borer and leaf hopper-plant resistance in rice varieties. *Ent. Exp. Appl.* **12.** 789–800.

Poehlman, J. M. (1959). *Breeding field crops.* Holt, Rinehart and Winston, N.Y., pp. 427.

Chapter 14   *Insecticides and their Application*

The materials which are poisonous to insects can be classified as follows:

(a) *contact insecticides* which kill by external contact. They are absorbed through the cuticle;

(b) *internal insecticides* (stomach poisons) which must be ingested to be effective;

(c) *combined* internal and contact insecticides;

(d) *systemic insecticides* which are absorbed into plant tissues, either through the roots or the leaves, are transported within the plant and act eventually as internal insecticides against sap-sucking insects;

(e) *fumigants* which act in the gas phase and enter through the respiratory system and the cuticle.

The insecticides may be short-lived or residual. Some of the organophosphorus insecticides are active for only an hour or two before they are hydrolysed and rendered ineffective (e.g. TEPP). On the other hand some of the residual insecticides may remain effective for years (e.g. dieldrin in soil), and there are all grades in between.

The contact insecticides are used mainly against insects with sucking mouthparts. They can also be used against mandibulate insects whereas internal insecticides applied to the outside of plants are ineffective against sap suckers. Internal insecticides are sometimes mixed with attractant or food materials (such as bran) and used as baits.

Combined internal and contact insecticides such as the chlorinated hydrocarbons are highly effective insecticides which can be applied to the plant before it is attacked and remain effective for several weeks. These broad spectrum insecticides also kill many beneficial insects and may have deleterious effects on biological control. There is a move today towards more selective insecticides such as the systemics which kill sap suckers but not the beneficial insects living on the leaves of the host. As a general principal, insecticides should be used as sparingly as possible and should be as selective as possible and should offer minimum hazard to the health of man, livestock and beneficial insects. This is not easily achieved.

Prior to World War II there were a limited number of insecticides of moderate effectiveness, e.g. the internal poison included the arsenicals lead arsenate and Paris Green. Lead arsenate was widely used on leaf crops for caterpillar control but could not be used close to harvest because of its toxicity to man. It was used extensively on apples for control of codlin moth.

Paris green, a copper and arsenic compound was never used on plants but was used quite extensively for controlling mosquito larvae and in baits for grasshoppers and cutworms. Metaldehyde is still widely used in a bait for control of slugs and snails.

Among the contact insecticides a number of compounds were used, including:

(a) petroleum oils emulsified with water and used as sprays for the control of scale insects. They have both contact and internal action and are still used

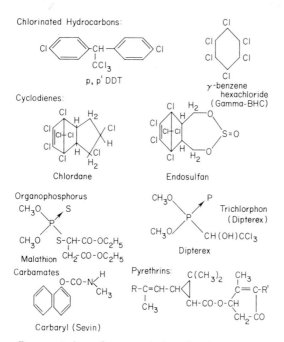

Representatives of some typical modern insecticides.

today in combination with other insecticides though they are only moderately effective when used alone;

(b) coal tar derivatives, e.g. DNOC (dinitro-orthocresol) against the eggs of red spider mites and aphids. Many of these compounds are very hazardous to man and are not used today;

(c) nicotine sulphate was for many years the most effective contact insecticidal spray and glasshouse fumigant for the control of aphids and mites;

(d) sulphur—especially applied as a very fine dust (micronized)—is very effective against some mites;

(e) pyrethrins, which are still very important insecticides, are derived from extract of flowers of the pyrethrum plant (*Chrysanthemum cinerariaefolium*):

Several chemically related substances are involved and these penetrate the cuticle and act rapidly on the central nervous system to knock down insects. They are important constituents of domestic fly sprays because of their excellent knock down action and low human toxicity. Their activity can be increased by the addition of synergists such as piperonyl butoxide. Other insecticides are often mixed with pyrethrum as the knock down effect may be of short duration;

(f) rotenoids. Extracts of derris roots have been used for insect control for over 100 years and they have been used very much longer in primitive communities as fish poisons. The most important active constituent is rotenone which occurs in 68 species of plants. Although it is very toxic to some insects it is harmless to plants and of low toxicity to man. It is often applied in a dust and enters the insect body through the alimentary canal, the integument or the tracheal system. It inhibits the respiratory enzyme glutamic acid oxidaze and thus suffocates the insect.

During World War II the insecticidal properties of DDT were discovered. The chemical had been known since 1874 but its biological properties had been unknown. It proved to be extraordinarily toxic to a wide variety of insects at very low concentrations. One of its first large scale uses was in Naples in 1945 where a typhus epidemic was arrested by treating all the population with DDT dust to kill their lice. This was a great step forward in medical entomology and as DDT was also found to be highly effective against flies and mosquitoes it was hailed as having great potential for public health use. Indeed it is still used in large amounts in malaria control and eradication programmes. As more DDT became available it was used extensively in agriculture and rapidly replaced many of the older insecticides.

The chemists, stimulated by the success of DDT, began to search for other synthetic organic insecticides. Another chlorinated hydrocarbon with high insecticidal activity was the gamma isomer of benzene hexachloride (Gamma BHC). This was more toxic towards mites and ticks. In the late 1940s another major group of synthetic organic insecticides was developed in Germany. These were the organophosphorus compounds, some of which were very effective against plant parasitic mites and aphids which were not affected by DDT, e.g. parathion, TEPP. Unfortunately these compounds were also extremely poisonous to man (e.g. TEPP is four times as toxic as strychnine to mammals). Today they have largely been replaced by less toxic organophosphorus insecticides such as diazinon, malathion and trichlorphon.

One group of organophosphorus compounds were particularly interesting in that they were taken up through the roots or shoots of plants and transported in the vascular system and were therefore selectively poisonous to sap suckers but not to their natural enemies. Schradan was the first of these compounds but it too was very poisonous to man. Some less toxic systemics have since been developed, e.g. demeton-methyl and dimethoate. Mevinphos, though highly toxic, is rapidly broken down in the plant and can be used closer to harvest.

Some highly toxic systemics of low solubility have been used as soil treatments to protect newly planted seeds and young seedlings, e.g. phorate, disulfoton. Some of the low toxicity organophosphorus compounds have been used against insect parasites of livestock, e.g. trichlorphon, fenchlorphos.

Another chemically distinct group of insecticides are the cyclodienes, e.g. chlordane, heptachlor, aldrin and dieldrin. These are not much used today because of their human toxicity or (in the case of aldrin and dieldrin) because of their very long persistence in the soil.

The carbamates, e.g. carbaryl, are a recently developed family of insecticides which show promise because of their low toxicity to mammals and rapid elimination from the body.

Following the widespread use of chlorinated hydrocarbon insecticides the red spider mites became more serious pests and a number of new acaricides have been developed to deal with these pests, e.g. chlorfenethol, dicofol, tetradifon and others. Some of these are active against all stages of the mites while others are more active against eggs. For the most part they have little insecticidal activity.

There are many ways of testing insecticides (see Busvine, 1971). For statistical reasons it is usual to refer to relative toxicity as the LD_{50} for each compound. This is the amount of insecticide (expressed in micrograms/gram or mg/kg body weight) which is required to kill 50% of the test population. Thus a low LD_{50} implies a highly toxic compound while a higher LD_{50} indicates a less toxic substance. The values of L_{50} will differ whether one is administering the material externally or internally and whether a single large dose is used or frequent smaller doses. Where a single dose is administered by mouth we refer to the acute oral toxicity. In order to get some comparison of relative mammalian toxicity of some commoner insecticides the acute oral LD_{50} values found (mainly with rats) are given in Table 18 (see pp. 138–9).

Some of these substances, say with LD_{50} below about 20 mg/kg, are obviously very toxic to mammals and therefore spray operators, and should not be used in the tropics except by skilled personnel with proper protective clothing. Martin (1969) gives a useful summary of the code of conduct drawn up for pesticide users by a committee representing British manufacturers, users and distributers of pesticides as well as conservation organizations. This code points out the need for a proper identification of the problem and the use of appropriate chemicals and methods of application to ensure safety to operators and others. It should be studied by anyone who proposes to use pesticides or to employ others to use them. Operators must be properly trained.

Table 18. Relative mammalian toxicity of some insecticides and acaricides

Name	LD_{50} (mg pesticide/ kg body weight)	Test animal	Type of action
Azinphos-methyl III	16·4	Rat	Insects, mites, long persistence
Gamma-BHC	88–91	Rat	Stomach and contact
Carbaryl	850	Rat	Contact and stomach; toxic to fish, earthworms
Chlorbenside	10,000	Rat	Acaricide
Chlordane	457–590	Rat	Stomach and contact
Chlorfenvinphos III	10–39 117–200	Rat, Mouse	
Chlorobenzilate	700–3,100	Rat	Acaricide
DDT	113	Rat	Stomach and contact
Demeton-methyl III	40–180	Rat	Systemic
Diazinon	150–600	Rat	
Dichlorvos III	56–80	Rat	Contact, stomach, fumigant
Dieldrin	46	Rat	Stomach—very persistent in soil
Dimethoate	500–600	Rat	Systemic and contact
DNOC II	25–40	Rat	Stomach, contact
Endosulfan II	55–220	Rat	Stomach, contact; very toxic to fish
Fenchlorphos	1,740	Rat	Systemic and contact
Fenitrothion	250–500	Rat	Contact; toxic to fish
Isobornyl thiocyano-acetate	1,603	Rat	Contact, knockdown
Heptachlor	100–162	Rat	Contact and stomach
Malathion	2,800	Rat	Contact, insecticide and acaricide
Menazon	1,950	Rat	Systemic
Metaldehyde	600–1,000	Dog	Molluscicide baits
Methiocarb	100 40	Rat, Guinea pig	Molluscicide baits
Methoxychlor	6,000	Rat	Stomach and contact
Mevinphos II	3·7–12	Rat	Contact, short persistence

Table 18 (*contd.*)

Name	LD_{50} (mg pesticide/ kg body weight)	Test animal	Type of action
Monocrotophos	13–23	Rat	Systemic, contact; toxic to bees, fish
Nicotine III	50–60	Rat	Contact
Parathion II	3·6–13	Rat	Contact
Paris green	22	Rat	Stomach, human skin
Phorate II	1·6–3·7	Rat	Systemic
Phosalone	120–170	Rat	Insecticide and acaricide
Phosmet	230–299	Rat	Insecticide and acaricide
Phosphamidon III	28	Rat	Systemic
Phoxim	2,000 375	Rat, Rabbit	Stored products and domestic use
Pirimicarb	147	Rat	Systemic in xylem
Pirimiphos-methyl	2,080 30–60	Rat, Fowl	Contact and fumigant, insecticide and acaricide
Proxopur	100	Rat	Contact
Pyrethrins	200	Rat	Contact
Rotenone	132–1,500	Rat	Contact; toxic to pigs and fish
Ryania	750–1,000	Rat	Stomach
Schradan II	9–42	Rat	Systemic
Sodium fluoride	75–150	Man	Stomach
Sodium fluoracetate	0.22	Rat	Intense mammalian poison (1080)
TDE	3,400	Rat	Contact and stomach
TEPP II	1·12 2·4 Dermal	Rat Rat	Contact; rapidly hydrolysed by water
Tetrachlorvinphos	4,000–5,000	Rat	Selective
Tetradifon	5,000	Rat	Acaricide
Trichlorphon	560–630	Rat	Contact and stomach
Vamidothion III	100–105	Rat	Systemic insecticide and acaricide

(NOTE: Terminology and LD_{50} after Martin, 1971; substance marked II or III are scheduled under the 1966–67 Regulations of the British Agriculture (Poisonous substances) Act. 1952 for restricted use or requiring special handling precautions.)

APPLICATION

Pesticides may be applied as dusts, sprays or fumigants. In dusts the active insecticide is mixed with a powdered, inactive carrier and is applied with a hand or power operated crop duster. Although simple and inexpensive hand dusters are widely used in developing countries, this is the least efficient method of application since it is difficult to get good crop coverage and the dust may create a health hazard. Granular formulations are sometimes used, especially for applying organophosphorus insecticides to soil or bodies of water where a gradual and continuing release of active material is required. These are safer to handle than liquid formulations.

The majority of pesticides are applied as sprays in which the active material is added to a liquid carrier (usually water but sometimes oil) and the mixture is broken up into droplets of spray which are applied to the crop. For efficient spraying a sufficient (but not excessive) amount of active insecticide must be uniformly deposited on the insects or on their food plants. One of the most important factors in ensuring efficient spray coverage is the production of spray droplets of the right size range. In general, the smaller the droplet size, the more efficient the spray coverage (down to droplets of diameter about $30\mu = 30 \times 10^{-6}$ m). Below this size there tends to be excessive loss by evaporation or spray drift.

Smaller particlesof in secticide (10–$30/\mu$) are used as aerosol fogs of smokes which are usually directed against flying insects since the particles settle very slowly and drift readily at low wind speeds. The carrier is usually compressed air or inert gas instead of water, or the insecticide may be volatilized by heat if it is thermostable.

The smallest particles of insecticides are of molecular dimensions in the gas phase of fumigants. These are normally used in an enclosed space so that a toxic concentration can be maintained for sufficient time, e.g. in granaries, houses, ship's holds or under gas-tight tents). Penetration may be a problem and this is often improved by vacuum fumigation where part of the air is first withdrawn and replaced by fumigant. As with all other insecticides the efficiency of different fumigants vary from one insect species to another. Some fumigants may be adsorbed by the material treated resulting in lower insecticidal efficiency or tainting of the product. Even very small quantities of some insecticides impart an objectionable taste to some foods. Larger quantities of insecticides in food may constitute a health hazard and many countries have laws regulating the maximum amounts of various insecticides permissible in foods.

Pesticides are usually sold commercially in a form which requires to be mixed with an appropriate volume of water or other diluent before use. It may be in solution, as a wettable powder or an emulsifiable concentrate. In any case the makers recommendations should be studied carefully and great care taken in handling the concentrated material which is usually highly toxic and potentially dangerous. Protective clothing is often essential at this stage.

The volume of spray applied per unit area will vary according to the type of crop, the nature of the spray equipment and the size of spray particles required:

	Gallons/acre	
	Tree crops	Ground crops
Ultra low volume (ULV)	<1	<1
Very low volume	1–5	1–5
Low volume	20–50	5–20
Medium volume	50–100	20–60
High volume	>100	>60

(NOTE: 100 gal/acre = 110 l/hectare).

The lower the spray volume, the more concentrated the spray. In the case of ultra low volume the concentrated insecticide mixed with an oil base is applied at the rate of a pint or two per acre. As smaller spray volumes are used the spray becomes more potentially toxic to the operator and stringent precautions must be taken for protection. However, the great advantage of this method is that it eliminates the need for carting large volumes of water in plantations and the equipment for application is often much simpler and cheaper.

There is a great variety of types of spray equipment. Portable, hand-operated knapsack sprayers operating at 20–80 p.s.i. and carrying 1–3 gallons are widely used in the tropics. In recent years there has been increasing use of motorized knapsack sprayers with engines creating a blast of air into which the insecticide is injected at low pressure through an air-shear nozzle. Large plantations may have tractor drawn ground crop or tree crop sprayers of 40–100 gallon capacity operating with hydraulic nozzles or using air blast and air shear nozzles. In any case the equipment must be properly adjusted to deliver the required spray droplet size and to ensure that the insecticide is properly mixed with the carrier.

There has been a considerable use of fixed wing aircraft and helicopters for crop spraying, usually at 110 l/Ha or 10 l/Ha. These tend to be less effective for trees since most of the spray is deposited on the upper surface of leaves at the tops of the trees and on the windward side. Aircraft have also been used extensively for locust control.

It is most important that consideration be given to the effects of insecticides on human health, wildlife and beneficial insects before programmes are undertaken. The chosen insecticide should be specific and applied at the right dose rate at the right time and used with due regard to the safety of the operator and others.

FURTHER READING

Busvine, J. R. (1971). *Techniques for testing insecticides.* 2nd Ed. Commonwealth Agricultural Bureau, London, pp. 345.

Clayphon, J. E. and Matthews, G. A. (1973). Care and maintenance of spray equipment in the tropics. *PANS* **19**(1). 13–23.

Gunther, F. A. and Jeppson, L. R. (1960). *Modern insecticides and world food production.* London, pp. 284.

Jacobson, M. and Crosby, D. G. (Ed.) (1971). *Naturally occurring insecticides.* Marcel Dekker, N.Y., pp. 585.

Martin, H. (Ed.) (1969). *Insecticide and fungicide handbook.* Blackwells, Oxford, pp. 387.

Martin, H. (Ed.) (1971). *Pesticide manual.* 2nd. Ed. British Crop Protection Council, pp. 495.

Matthews, G. A. and Clayphon, J. E. (1973). Safety precautions for pesticide applications. *PANS* **19**(1). 1–12.

Morgan, W. G. (1972). Spray application in plantation crops. *PANS* **18**(3). 316.

Ripper, W. E. (1955). Application methods of crop protection chemicals. *Ann. Appl. Biol.* **22.** 288–324.

Ripper, W. E. and George, L. (1965). *Cotton pests of the Sudan.* Blackwells, Oxford, pp. 345

Chapter 15   *Malaria Control and Eradication*

Malaria is the most widespread and damaging disease of many tropical developing countries. It causes high mortality, especially in the first two years of life, and its chronic recurrence with severe anaemia and general debility causes much sickness and diminished efficiency at work. Today some 1,842 million people live in 145 malarious countries.

The protozoan malaria parasite multiplies asexually in man in the blood and blood-forming organs. It is transmitted by mosquitoes of the genus *Anopheles* and undergoes sexual development in the vector, a process which takes about 12 days following a meal of infected blood. After this period the mosquito is capable of transmitting the parasites in its saliva when feeding.

Control methods are aimed at breaking the transmission cycle: by reducing contacts between man and mosquitoes, by reduction of mosquito populations and by drug treatment of human infections. These methods require an efficient public health service and the education of populations at risk.

Contacts between man and mosquitoes are reduced by the siting and screening of houses or sleeping places and by the use of suitable clothing and repellents after dusk when most malarial mosquitoes become active. Mosquito populations can be reduced by elimination of breeding places close to houses and by larval control where breeding places cannot be destroyed or drained. Insecticides can be used either against larvae or space sprays or residual sprays can be used against adults. Suppressant drugs are widely used in malarial areas and recognized cases can be cured by appropriate drug treatment.

Prior to World War II quinine was the only effective drug readily available and control was mainly aimed at larval populations using Paris Green or oils as larvicides and the elimination of breeding sites by drainage and other engineering methods. Where sufficient funds and effort were available these control methods had some local success. Indeed some large scale projects such as the draining of the Pontine Marshes around Rome were highly effective. There were however no satisfactory residual insecticides available for use against adults and the contact space sprays based on pyrethrum were very expensive.

The discovery of the powerful insecticidal activity of DDT and its long-term residual effects in the late 1940s was of immediate interest to public health workers. In the next decade its use as residual deposits on wall surfaces was dramatically effective in interrupting malaria transmission in endemic areas. Thus it was found that malaria could be eradicated without totally eradicating the vector (an expensive and often impracticable undertaking).

In 1955 the World Health Assembly instructed the World Health Organization to prepare and implement a programme for malaria eradication. In 1956 the WHO expert committee on malaria laid down the principles and practice of such a programme which was defined as ". . . the ending of transmission of malaria and the elimination of the reservoir of infective cases in a campaign limited in time and carried to such a degree of perfection that, when it comes to an end, there is no resumption of transmission."

The WHO programme for eradication extends over 8–10 years and takes place in four phases:

(1) the *preparatory phase* involving epidemiological and geographical reconnaisance, the establishment of services and staff training;

(2) the *attack phase* involving the thorough spraying of all houses in the operational area with DDT;

(3) the *consolidation phase* where there is active and intense surveillance to eliminate remaining infections by drug treatment and to prove the eradication of malaria;

(4) the *maintenance phase* which continues after malaria has been eradicated in the operational area and continues until world-wide eradication has been effected.

The rationale behind this programme depends upon the behaviour of the anopheline mosquito and the parasite life history. Most transmission occurs at night in sleeping quarters. After a mosquito takes an infected blood meal it takes about twelve days for completion of the parasite's sexual cycle and then the mosquito becomes infectious. Assuming that a mosquito feeds every two days, it will return to a house six times before it becomes infectious. If the interior walls of *all* houses have been sprayed with DDT it is highly probable that the mosquito will be killed before it becomes infectious. Under these conditions the transmission of malaria will cease and no new cases will appear. This is the central theme of malaria eradication.

At first entomologists sprayed houses with DDT every year but when a shortage of DDT occurred in Greece in 1951, spraying which had been going on for five years in some areas was stopped. No new cases appeared and it was concluded that if transmission had been fully prevented for some years, spraying could cease. It is important that medical teams should keep a careful watch for new cases (perhaps coming in from outside) and treat them immediately. In about three years most species of malaria parasites are eliminated from the blood—even without treatment.

The early WHO recommendation was for a "time limited campaign" and this is important because DDT-resistant strains of mosquitoes are progressively selected in long-term programmes.

By 1970 eradication had been effected in 37 countries, giving protection to 39% of the people living in malarial areas. A further 50 countries had embarked on eradication programmes and 38 more were engaged in other antimalarial programmes. The result has been an enormous decline in morbidity. Although the WHO eradication programme has been dramatically effective there have been failures in some countries and these can be attributed to a variety of causes. There are enormous problems in planning and organizing such a large scale operation, particularly where technical, financial or education resources are limited.

Resistance to DDT and other insecticides is now widespread. In some areas the use of residual insecticides has been unsuccessful because of outdoor feeding of mosquitoes or the practices of some people of sleeping outdoors or under rudimentary shelters. Human ecology must be considered as well as mosquito ecology; indeed this is of vital importance.

Although some of the newer insecticides are just as effective as DDT they are from three to nine times as expensive and in some cases are more hazardous for the operator to apply. Since DDT for malaria eradication now costs 60 million US dollars per annum, this cost factor is obviously important.

Where eradication with residual insecticides has proved impracticable it is necessary to adopt a suitable control strategy which takes into account the local ecological situation as well as the socio-economic priorities of the people concerned. Over the last two decades the emphasis has been mainly on the use of insecticides but there are indications of a move towards a more rational integration approach using biological control and environmental sanitation in conjunction with insecticides and the use of drugs. Predacious fish such as *Poecilia* and *Gambusia* have long been used and some current research is investigating the use of insect pathogens and mosquito control by genetic measures.

The entomologist has a vital role in antimalarial operations. He must identify the significant mosquito vectors and study their ecology and feeding behaviour in the operational area. Later he must be able to supervise spraying operations and ensure that the proper amounts of insecticide are applied. He must be alert to the appearance of insecticide resistance and should be aware of the effects of insecticides on other members of the ecosystem in order to avoid undesirable side effects and environmental pollution. He must work closely with the parasitologist and other members of the malaria eradication team.

FURTHER READING

Macdonald, G. (1957). *The epidemiology and control of malaria*. London.
Mattingly, P. F. (1962). Mosquito behaviour in relation to disease eradication programmes. *Ann. Rev. Ent.* **7.** 419–436.

Mattingly, P. F. (1969). *The biology of mosquito borne disease.* Allen and Unwin, London, pp. 184.

Pampana, E. J. (1969). *A textbook of malaria eradication.* 2nd. Ed., Oxford University Press, London, pp. 187.

Wright, J. W., Fritz, R. F. and Haworth, J. (1972). Changing concepts of vector control in malaria eradication. *Ann. Rev. Ent.* **17**. 75–102.

Chapter 16   *Summary of Major Insect Pests of Tropical Crops*

Some major pests of tropical crops are listed in this chapter (Tables 20–26) and some particularly serious or widespread pests are indicated by an asterisk (*). This list is selective as large numbers of insects have been recorded on some crops (e.g. 4,098 species on rice alone).

The major tropical crops are listed in Table 19 in terms of area and production. This information alone does not give a sufficient picture of their importance as some crops are much more valuable per unit weight (e.g. tea, coffee, cocoa), while others are of vital importance for subsistence and are only secondarily exported (rice, sweet potatoes, yams, bananas, coconuts). In South East Asia and the Pacific the coconut palm provides food, fibres and building materials. Rice is by far the most important subsistence crop in the tropics.

Table 19. Production of major tropical crops in 1971

Crop	Area cultivated (millions of hectares)		Production (millions of metric tons)	
	World	Tropics	World	Tropics
Rice	134·9	100·0	307·4	196·3
Millet and sorghum	113·4	73·7	101·1	51·8
Wheat	217·2	68·2	343·1	80·6
Maize	112·9	60·2	307·8	82·1
Oilseeds	120·8	55·5	112·0	27·8
Pulses	62·5	42·6	44·8	20·2
Cotton	33·1	20·1	11·8	5·3
Groundnuts	18·8	17·0	18·5	17·1
Sugar cane	11·0	10·9	83·6	50·4
Cassava	9·8	9·8	92·2	92·2
Sweet potato and yams	17·0	5·4	147·7	43·8
Bananas	1·8	1·8	28·2	27·9
Coffee	—	—	4·9	4·9
Tea	1·3	1·3	1·3	1·3
Cocoa	—	—	1·5	1·5
Rubber	—	—	3·0	3·0
Copra	—	—	3·7	3·7

(Data from FAO Production year book, **25**, 1971. Rome.)

Particular attention is given to agricultural pests because the bulk of the world's agricultural population is found in the tropics (see Fig. 30). The value of the main subsistence crops (especially rice) far exceeds that of the tropical cash crops grown for export (Fig. 31) and this should give some indication of where the efforts of economic entomologists might well be directed.

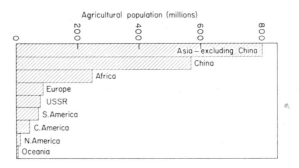

Fig. 30. World distribution of Agricultural population.
(Data from *FAO Production Year Book* **25**, 1971.)

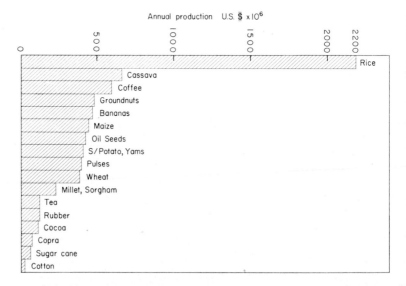

Fig. 31. Approximate relative value of annual production of some major crops (in millions of US dollars) in the tropics. (Data from FAO Production Year Book and FAO Trade Year Book **25**, 1971.)

Table 20. The major pests of coffee and their distribution
(most important pests marked with an asterisk*)

	Method of attack	Distribution
ORDER: ACARINA		
FAMILY: TETRANYCHIDAE		
Oligonychus coffeae (Niet.)	Leaves	Africa, India, Indonesia, Asia, Ceylon
ORDER: HEMIPTERA		
FAMILY: PSEUDOCOCCIDAE		
Planococcus citri (Risso)	Roots, leaves, fruit	Pantropical
P. kenyae (LePelley)	Leaves, shoots, berries	Africa
P. lilacinus (Ckll.)	Leaves, shoots, berries	South East Asia, India, Philippines, Indonesia
Pseudococcus longispinus (Tar.)	Leaves. shoots	Ceylon, Vietnam, Indonesia
Ferrisia virgata (Ckll.)	Leaves, shoots, berries	Africa, Pacific, New Guinea, Philippines
Nipaecoccus vastator (Mask.)	Leaves, shoots	Africa, Madagascar, Java, Philippines
Chavesia caldasiae Bal.	Roots	Colombia
Rhizoecus spp.	Roots	South America
Neorhizoecus spp.	Roots	South America
Paraputo leveri (Green)	Roots	New Guinea
Puto spp.	Roots	South America
Geococcus coffeae Green	Roots	South America
FAMILY: COCCIDAE		
Coccus viridis (Green) (Green scale)	Leaves, shoots, berries	Pantropical
Gascardia brevicauda (Hall)	Shoots, twigs	Africa
Saissetia coffeae (Walk.)	Leaves, branches, roots, fruit	Pantropical

contd.

Table 20 (*contd.*)

	Method of attack	Distribution
FAMILY: COCCIDAE (*contd.*)		
Parasaissetia nigra (Niet.)	Leaves, shoots, branches	Central and South America, Africa, India, Malaysia, Pacific, New Guinea
Pulvinaria psidii Mask.	Leaves, shoots	Africa, India, Indonesia, New Guinea, Pacific, Central America
FAMILY: ASTEROLECANIIDAE		
Asterolecanium coffeae Newst.	Trunk, shoots	Africa
Cerococcus catenarius da Fons.	Young branches	Brazil
FAMILY: DIASPIDIDAE		
Selenaspidus articulatus (Morg.)	Leaves, fruit	South America
FAMILY: TINGIDAE		
Habrochila spp.	Leaves (defoliating)	Africa
Dulunius unicolor (Sign.)	Leaves	Madagascar
FAMILY: MIRIDAE		
Lamprocapsidea coffeae (Chin.)	Flowers	Africa
FAMILY: APHIDIDAE		
Toxoptera aurantii B. de F.	Leaves, shoots	Pantropical
FAMILY: PENTATOMIDAE		
Antestiopsis spp.	Green berries, buds	Africa, India, Malaysia
ORDER: THYSANOPTERA		
Diarthrothrips coffeae Will.	Leaves, shoots, berries	Africa
Scirtothrips sweetmanni Bianch.	Leaves	India

Table 20 (*contd.*)

	Method of attack	Distribution
ORDER: LEPIDOPTERA		
FAMILY: COSSIDAE		
Zeuzera coffeae Niet.	Stem borer	India, Ceylon, Vietnam, Malaysia, Indonesia, New Guinea
Xyleutes spp.	Stem borers	Central and South America, Africa, Taiwan, Philippines
FAMILY: TORTRICIDAE		
Homona coffearia (Niet.)	Young fruit, leaves	India, Ceylon, Indonesia, New Guinea
Eucosma nereidopa Meyr.	Berries, shoots	Africa
FAMILY: LYONETIIDAE		
Leucoptera spp.	Leaf miners	Central and South America, Africa
FAMILY: LIMACODIDAE		
Parasa vivida (Wlk.)	Leaves	Africa
P. lepida (Cramer)	Leaves	South East Asia
Setora nitens (Wlk.)	Leaves	Vietnam, Indonesia, New Guinea
Cheiromettia spp.	Leaves	Indonesia
FAMILY: PYRALIDAE		
Dichocrocis crocodera (Meyr.)	Leaf roller	Africa
Prophantis smaragdina (Bail.)	Berries	Africa
FAMILY: LYCAENIDAE		
Virachola bimaculata (Hew.)	Berries	Africa
FAMILY: SPHINGIDAE		
Cephanodes hylas (L.)	Leaves	Africa, Vietnam, Malaysia

contd.

Table 20 (*contd.*)

	Method of attack	Distribution
FAMILY: NOCTUIDAE		
Achaea catacaloides Guen.	Leaves, army worm	Africa
Tiracola plagiata (Wlk.)	Leaves, fruit, army worm	India, Malaysia, Indonesia, New Guinea
Agrotis spp.	Cutworms	Africa, Indonesia, New Guinea
FAMILY: LYMANTRIIDAE		
Euproctis spp.	Leaves, bark	Africa, Asia, New Guinea
FAMILY: DREPANIDAE		
*Epicampoptera spp.	Leaves (defoliated)	Africa
Oreta extensa Wlk.	Leaves (defoliated)	Indonesia
FAMILY: GEOMETRIDAE		
Ascotis selenaria reciprocaria Wlk.	Leaves, fruit, flowers	East Africa
Epigynopteryx spp.	Leaves, fruit	Kenya
Hyposidra spp.	Leaves	Malaysia, Indonesia

ORDER   COLEOPTERA

FAMILY: CICINDELIDAE		
Collyris spp.	Stem borer	South East Asia
FAMILY: BOSTRYCHIDAE		
*Apate monachus F.	Stem borer	Africa, Central and South America

Table 20 (*contd.*)

	Method of attack	Distribution
FAMILY: CERAMBYCIDAE		
Xylotrechus spp.	Trunk borers	South East Asia
Anthores leuconotus Pasc.	Trunk borers	Africa
Bixadus sierricola (White)	Ring barking	Africa
Dirpha spp.	Ring barking, stem borers	Africa
FAMILY: CURCULIONIDAE		
Cryptorhynchus spp.	Larval ring barking	New Guinea
Meroleptus cinctor Mshl.	Larval ring barking	New Guinea
Diaprepes familicus (Oliv.)	Roots	West Indies
Lachnopus coffeae Mshl.	Shoots, flowers, fruits	Puerto Rico
Ellimenistes laesicollis F. (Scolytinae)	Natal	
Hypothenemus hampei (Ferr.)	Berry borer	Pantropical
Xylosandrus compactus (Eich)	Twig borer	Africa, South East Asia, Pacific
Xylosandrus morigerus (Bldf.)	Twig borer	Africa, South East Asia, Philippines, New Guinea, Australia, Pacific

ORDER: DIPTERA

FAMILY: CECIDOMYIDAE		
Contarinia coffeae Harris	Fruit	Uganda
FAMILY: TRYPETIDAE		
Ceratitis capitata (Wied.) (Mediterranean fruit fly)	Fruits	Mediterranean

Table 21. The major pests of tea and their distribution
(most important pests marked with an asterisk*)

	Method of attack	Distribution
ORDER: ACARINA		
FAMILY: TETRANYCHIDAE		
Oligonychus coffeae (Niet.)	Defoliates	North India
Brevipalpus californicus (Banks)	Leaves	South India
B. phoenicis (Geij.)	Leaves	North India
B. obovatus Donn.	Leaves	Indonesia, Ceylon, Japan
FAMILY: ERIOPHYIDAE		
Calacarus carinatus (Green)	Leaves, depressed growth	South India
Acaphylla theae (Watt)	Leaves	South India
FAMILY: TARSONEMIDAE		
Hemitarsonemus latus (Banks)	Leaves, polyphagous	Widespread
ORDER: ISOPTERA		
FAMILY: KALOTERMITIDAE		
Neotermes militaris (Desn.)	Nest in stems	Ceylon
FAMILY: TERMITIDAE		
Microcerotermes spp.	Nest in soil	Assam
ORDER: THYSANOPTERA		
Scirtothrips bispinosus (Bagn.)	Leaves	South India
Taeniothrips setiventris (Bagn.)		Darjeeling
Scirtothrips dorsalis (Hood)		North India

Table 21 (*contd.*)

	Method of attack	Distribution

ORDER: HEMIPTERA

FAMILY: APHIDIDAE

Toxoptera aurantii B. de F.	Nursery pest	Widespread

FAMILY: CICADELLIDAE

Empoasca flavescens (F.)	Leaves	India
E. onukii Mats.	Leaves	Japan

FAMILY: COCCIDAE

Coccus viridis (Green)		Ceylon and South India
Saissetia coffeae (Wlk.)		Ceylon and South India
S. formicaria (Green)		India
Eriochiton theae (Green)		North India
Phenacaspis manri (Green)		North India
Ceroplastes rubens Mask.		North India
Fiorinia theae Green		North India

FAMILY: MIRIDAE

Helopeltis antonii Sign.	Buds, young leaves	India, East Asia, Malaysia, Indonesia
H. clavifer Walk.		New Guinea
H. bergrothi Reut.	Stem canker	Africa
*H. theivora Waterh.	Buds, young leaves	India, Ceylon

ORDER: LEPIDOPTERA

FAMILY: COSSIDAE

Zeuzera coffeae (Niet.)	Larval stem borer	

contd.

Table 21 (*contd.*)

	Method of attack	Distribution
FAMILY: GRACILLARIIDAE		
Gracillaria theivora Wals.	Leaf miner and roller	
FAMILY: LIMACODIDAE		
Setora nitens (Wlk.)	Leaves	South East Asia
Parasa lepida (Cramer)	Leaves	Ceylon
Thosea spp.	Leaves	India, Ceylon
Macroplectra nararia (Moore)	Severe damage	India, Ceylon
FAMILY: NOTODONTIDAE		
Stauropus alternus Wlk.		
FAMILY: BOMBYCIDAE		
Andraca bipunctata Wlk.	Bunch caterpillar	
FAMILY: GEOMETRIDAE		
Buzura suppressaria (Guen.)	Leaves	India
FAMILY: PSYCHIDAE		
Mahasena theivora (Dudg.)		
Acanthopsyche subteralbata (Hamp.)		
Clania spp.		
Manatha spp.		
FAMILY: TORTRICIDAE		
*Homona coffearia (Niet.)	Tea tortrix, leaf roller	India, Indonesia

Table 21 (*contd.*)

	Method of attack	Distribution
FAMILY; TORTRICIDAE (*contd.*)		
Cydia leucostoma (Meyr.)	Flushworm, leaf roller, buds and upper leaves	India, Indonesia
Adoxophyes orana (Roels.)	Leaf roller	Japan
Caloptilia theivora Wals.	Leaf roller	Japan
FAMILY: NOCTUIDAE		
Spodoptera litura (F.)	Army worm	
Tiracola plagiata (Wlk.)	Army worm	
ORDER: COLEOPTERA		
FAMILY: SCARABAEIDAE		
Holotrichia spp.	Larvae eat roots and girdle collar	India, Ceylon
Exopholis hypoleuca Wied.	Roots	Indonesia
Anomala superflua Ar.	Leaves	
FAMILY: CERAMBYCIDAE		
Aolesthes induta Newm.		New Guinea
FAMILY: CURCULIONIDAE		
Dicastigus mlangensis Mshl.	Adults on leaves	East Africa
Hypomeces squamosus Herbst.	Adults on leaves	Indonesia
Xylosandrus fornicatus Eich.	Tea shot hole borer	
Xyleborus compactus (Eich.)		Japan
X. germanus (Bldf.)		Japan
X. villosus Schedl		Argentina

Table 22. The major pests of cotton and their distribution
(most important pests marked with an asterisk*)

	Method of attack	Distribution
ORDER: ISOPTERA		
FAMILY: HODOTERMITIDAE		
Hodotermes mossambicus (Hag.)	Seedlings	East Africa
FAMILY: TERMITIDAE		
Microtermes thoracalis Sjost. (Cotton soil termite)	Roots	Africa
ORDER: ORTHOPTERA		
Schistocerca gregaria Forsk. (Desert locust)		
Locusta migratoria migratorioides R. and F. (Migratory locust)		
Nomadacris septemfasciata Serv. (Red locust)		
ORDER: HEMIPTERA		
FAMILY: CICADELLIDAE		
Empoasca lybica (de B.)	Leaf scorch	Sudan
E. facialis (Jac.)	Leaf scorch	South, Equatorial and West Africa
E. devastans Dist.	Leaf scorch	India
Erythroneura lubiae (China)	Leaf spotting	Sudan

Table 22 (*contd.*)

	Method of attack	Distribution
FAMILY: APHIDIDAE		
Aphis gossypii Glover.	Leaves. Stunting, boll shedding and vector of cotton anthocyanosis virus in Brazil	Cosmopolitan
FAMILY: ALEYRODIDAE		
Bemisia tabaci (Genn.)	Vector of cotton leaf curl virus	Africa
FAMILY: PSEUDOCOCCIDAE		
Maconellicocus hirsutus (Green)	Terminal shoots	Widespread
Ferrisiana virgata (Ckll.)		Widespread
Planococcus kenyae (Le Pelley)		Kenya
Nipaecoccus vastator (Mask.)		Africa
FAMILY: MARGARODIDAE		
Vrydagha lepesmei Vayss.	Roots	Africa
FAMILY: COCCIDAE		
Pulvinaria jacksoni Newst.	Leaves, shoots	Africa
Parasaissetia nigra (Niet.)		Africa
FAMILY: DIASPIDIDAE		
Pinnaspis strachani (Cool.)		Africa
FAMILY: MIRIDAE		
Helopeltis schoutedeni Reut. (Cotton mosquito bug)	Shoots, bolls	Africa
Creontiades pallidus (Ramb.)	Flower buds, young bolls	Africa
Lygus vosseleri Popp.	Leaf and flower buds	Africa

contd.

Table 22 (*contd.*)

	Method of attack	Distribution
FAMILY: LYGAEIDAE		
Oxycarenus hyalinipennis (Costa)	Seeds	Africa, India, Asia, Philippines, Brazil
Oxycarenus spp.		Six from Africa
Oncopeltus sordidus (Dall.)		Queensland
FAMILY: PYRRHOCORIDAE		
Dysdercus fasciatus Sign.	Cotton stainer	Africa
D. superstitiosus (F.)	Cotton stainer	Sudan
D. sidae Montr.		Australia, New Guinea, Pacific
D. singulatus (F.) (Red cotton bug)		India, Asia, Australia, New Guinea, Pacific
FAMILY: SCUTELLARIDAE		
Tectoris diophthalmus (Thunb.) Bolls		Queensland
FAMILY: PENTATOMIDAE		
Nezara viridula (L.)	Green bolls	Cosmopolitan
Calidea spp.	Green bolls	Africa

ORDER: THYSANOPTERA

Caliothrips impurus (Pr.) (Dark cotton leaf thrips)		Sudan

ORDER: COLEOPTERA

FAMILY: BUPRESTIDAE		
Sphenoptera spp.	Stem borers	India, Africa

Table 22 (*contd.*)

	Method of attack	Distribution
FAMILY: CHRYSOMELIDAE		
*Podagrica puncticollis Weise (Cotton flea beetle)		Africa
*P. pallida (Jac.) (Hambuk flea beetle)		Africa
FAMILY: APIONIDAE		
Apion soleatum Wagn.	Stem borer	Africa
A. varium Wagn.	Stem borer	Africa
FAMILY: CURCULIONIDAE		
Alcidodes brevirostris (Boh.) (Cotton stem girdling weevil)	Stems	Africa
*Anthonomus grandis Boh. (Cotton boll weevil)	Larvae and adults attack flowers and fruits	North and Central America, West Indies

ORDER: LEPIDOPTERA

	Method of attack	Distribution
FAMILY: LYONETIIDAE		
Bucculatrix gossypii Turn.	Leaf miner	Australia
FAMILY: GELECHIIDAE		
*Pectinophora gossypiella (Saund.) (Pink bollworm)	Bolls	Africa
P. malvella Hb.	Bollworm	North Africa, South East Asia
P. scutigera (Hold.)	Bollworm	Australia, New Guinea
P. eribdona Meyr.	Bollworm	Uganda

contd.

Table 22 (*contd.*)

	Method of attack	Distribution
FAMILY: PYRALIDAE		
Sylepta derogata F.	Leaf roller	Africa
FAMILY: SPHINGIDAE		
Hippotion celerio (L.)	Leaf worm	East Africa
FAMILY: NOCTUIDAE		
Diparopsis spp. (Red boll worm, Sudan boll worm)	Bolls	Africa
Earias biplaga Wlk.		Africa
E. vittella (F.) (Northern rough bollworm)		India, South East Asia, New Guinea, Australia
E. insulana (Boisd.) (Spiny bollworm)		Mediterranean, Africa, Asia (to West Malaysia)
Heliothis armigera (Hb.) (Corn earworm)		Cosmopolitan
H. virescens (F.)		Americas
H. zeae (Boddie)		Americas
Scadodes pyralis Dyar		South America
Alabama argillacea (Hb)	Leaf worm	Americas
Cosmophila flava (F.)	Leaf worm	Asia, Africa, Australia
Spodoptera litura (F.)	Leaf worm	Asia, New Guinea, Australasia, Pacific
S. exigua (Hb.)	Army worm	Africa, Asia, New Guinea, Australia, Pacific
S. littoralis (Boisd.) (Egyptian cotton worm)		Africa, Mediterranean
Agrotis spp.	Cutworms	Africa

Table 23. The major pests of cocoa and their distribution
(most important pests marked with an asterisk*)

	Method of attack	Distribution

ORDER: HEMIPTERA

FAMILY: PSEUDOCOCCIDAE

Planococcoides njalensis (Laing)	Main vector of swollen shoot virus	Africa
Planococcus citri (Risso)		Pantropical
P. lilacinus (Ckll.)		Madagascar, India, Ceylon, Malaysia, Philippines, Indonesia, Pacific
Pseudococcus adonidum (L.)		Cosmopolitan
Ferrisia virgata (Ckll.)		Pantropical

FAMILY: COCCIDAE

Howardia biclavis (Comst.)	Mining scale	Americas, Africa, India, Ceylon, Java, Australia

FAMILY: STICTOCOCCIDAE

Stictococcus aliberti Vauss.		Africa
S. sjostedti Ckll.		Africa

FAMILY: PSYLLIDAE

Tyora tessmanni (Aul.)	Leaf buds	Africa

contd.

Table 23 (*contd.*)

	Method of attack	Distribution
FAMILY: MIRIDAE*		
Sahlbergella singularis Hag.		Africa
Distantiella theobroma (Dist.)		Africa
Brycoropsis laticollis Schum.		Africa
Odoniella reuteri Hag.		Africa
Boxiopsis madagascariensis Lav.		Madagascar
Platyngomiriodes apiformis Ghauri		Malaysia
Pseudodoniella spp.		Papua New Guinea
Helopeltis clavifer Wlk. (approximately 12 species of Helopeltis on Old World cocoa)		New Guinea, Malaysia
Monalonion spp.		Central and South America
FAMILY: APHIDIDAE		
Toxoptera aurantii B. de F.	Young shoots	Cosmopolitan
Toxoptera theobromae Schout.		
FAMILY: CICADELLIDAE		
Empoasca devastans Dist.	Leaf burn	Africa, Ceylon
FAMILY: COREIDAE		
Amblypelta theobromae Brown.	Pod sucker	New Guinea
A. coccophaga China		Solomon Islands

Table 23 (*contd.*)

	Method of attack	Distribution

ORDER: LEPIDOPTERA

FAMILY: COSSIDAE

Zeuzera coffeae (Niet.)	Stem borer	Widespread
Eulophonotus myrmeleon Feld.	Stem borer	Africa

FAMILY: TORTRICIDAE

Cryptophlebia encarpa Meyr.	Pod borer	New Guinea

FAMILY: HEPIALIDAE

Endoclita hosei Tind.	Larvae boring and girdling stems	Malaysia

FAMILY: GRACILLARIDAE

Acrocercops cramerella Snell	Pod borer	Indonesia, Philippines

FAMILY: XYLORICTIDAE

Pansepta teleturga Meyr.	Bark web worm	New Guinea

FAMILY: NOCTUIDAE

Tiracola plagiata (Wlk.)	Army worm	New Guinea, Malaysia
Spodoptera litura F.	Leaf worm	Malaysia
Heliothis armigera (Hb.)	Leaf worm	Malaysia
Earias biplaga Wlk.	Buds, fruits	Africa
Anomis lena Schaus	Buds, fruits	Ghana

contd.

Table 23 (*contd.*)

	Method of attack	Distribution
ORDER: COLEOPTERA		
FAMILY: SCARABAEIDAE		
Adoretus versutus Har.	Roots and shoots	Madagascar, India, Pakistan, Java, Pacific
FAMILY: BUPRESTIDAE		
Chrysochroa bicolor F.	Stem borer	South East Asia
FAMILY: BOSTRYCHIDAE		
Apate monachus F.	Stem borer	West Indies, Africa
FAMILY: CERAMBYCIDAE		
Glenea novemguttata Geur.	Stem borer	Indonesia
G. aluensis Gah. and *G. lefeburei* Guer.		New Guinea
Mallodon downesi F.		Africa
Monochamus ruspator F.		Africa
Steirastoma breve Guby		West Indies, South America
Tragocephala nobilis F.		Africa
FAMILY: CURCULIONIDAE		
Pantorhytes spp.	Larval stem borers	New Guinea
Sphenophorus striatus F.	Larval stem borers	Fiji, San Thome
Xylosandrus compactus (Eich.)		India, Ceylon, South East Asia
Xyleborus discolor Bldf.		India, Ceylon, South East Asia
X. morstatti Hag.		Africa

Table 24. Major insect pests of sugar cane
(most important pests marked with an asterisk*)

	Method of attack	Distribution

ORDER: ISOPTERA

FAMILY: MASTOTERMITIDAE

Mastotermes darwiniensis Frogg.		Australia

FAMILY: TERMITIDAE

Odontotermes obesus Holmg.		India, Africa, West Indies, Australia

ORDER: HEMIPTERA

FAMILY: DELPHACIDAE

	Method of attack	Distribution
**Perkinsiella saccharicida* Kirk.	These are both vectors of the persistent sugar-cane Fiji disease virus	Australia, South East Asia, Hawaii, South America, Africa, Pacific
P. vastatrix Breddin		Philippines
Saccharosyne saccharivora (Westw.)		Central and South America

FAMILY: CERCOPIDAE

**Aeneolamia varia saccharina* Dist.		West Indies

FAMILY: CICADELLIDAE

Cicadulina mbila (Naude)	Vector of streak disease virus	Africa

FAMILY: APHIDIDAE

Sipha flava Forbes		West Indies, Central America

contd.

Table 24 (*contd.*)

	Method of attack	Distribution
FAMILY: APHIDIDAE (*contd.*)		
Rhopalosiphum maidis (Fitch)	Vector of sugar cane mosaic virus (also transmitted by 8 other species)	Cosmopolitan
FAMILY: PSEUDOCOCCIDAE		
Dysmicoccus brevipes (Ckll.)		Pantropical
Saccharicoccus sacchari (Ckll.) (Pink sugar cane mealybug)		Pantropical
FAMILY: COCCIDAE		
Aulacaspis tegalensis (Zhnt.)	Sugar cane scale	Africa, Malaysia, Java, Philippines

ORDER: LEPIDOPTERA

FAMILY: LYONETIIDAE		
Opogona glycyphaga Meyr.	Leaf miner	Australia
FAMILY: PYRALIDAE		
Chilo auricilius Dudg.	Sugar cane stem borer	India, Java, Taiwan
C. partellus (Swin.)		India, East Africa
C. infuscatellus Sn.		India, Burma, Philippines, Indonesia
Diatraea saccharalis (Fab.)	Oriental sugar cane borer	Americas
Diatraea sticticraspis (Hamp.)	Indian sugar cane borer	
Zeadiatraea lineolata (Wlk.)	Neotropical corn stalk borer	Central and South America
Proceras indicus Kapur	Internodal borer	India

Table 24 (*contd.*)

	Method of attack	Distribution
FAMILY: PYRALIDAE (*contd.*)		
P. sacchariphagus Bojer (Spotted borer)		Madagascar, Indonesia
Maliarpha separatella (Rag.)		Africa
Eldana saccharina Wlk.		Cosmopolitan
Emmalocera depressella Swin.	Cane root borer	India, Indonesia
Marasmia trapezalis Guen.	Leaf worm	West Indies, South America, Africa, Philippines
Omiodes accepta (Butl.)	Leaf roller	Hawai
FAMILY: NOCTUIDAE		
Sesamia inferens (Wlk.)	Violet rice stem borer	India, Java, Asia, New Guinea, Solomon Islands
S. botanephaga (T. and B.) and		
S. calamistis (Hamp.)		Africa
S. cretica Led.	Durra stem borer	Mediterranean, Africa
FAMILY: TORTRICIDAE		
Argyroploce schistaceana Sn.	Grey cane borer	Java, Philippines, Ceylon, Taiwan, Mauritius

ORDER: COLEOPTERA

FAMILY: SCARABAEIDAE		
Heteronychus arator (F.)	Roots	South Africa, Madagascar, Australia
Strategus spp.		Central and South America
Dermolepida albohirtum	Roots	Australia, New Guinea
Cochliotus sp.	Roots	Tanganyika

contd.

Table 24 (*contd.*)

	Method of attack	Distribution
FAMILY: SCARABAEIDAE (*contd.*)		
Ctenora smithi		West Indies
Eutheola fugiceps Lec.		USA
Lachnosterna spp.		West Indies, South America
Pseudoholophylla sp.		Queensland
FAMILY: ELATERIDAE		
Lacon variabilis Cand.	Roots	Queensland
FAMILY: CHRYSOMELIDAE		
Rhyparida morosa Jac.		Queensland
Dicladispa armigera (Oliv.)	Leaf miner	India, Ceylon, South East Asia, New Guinea
Asamangulia wakkeri (Zehnt.)		Indonesia
Rhadinosa lebongensis Maulik		India, China
FAMILY: CURCULIONIDAE		
*Rhabdoscelus obscurus (Boisd.)	Cane borer	New Guinea, Celebes, Pacific, Taiwan
Metamasius hemipterus (L.)	Rotten cane stalk borer	Central and South America, West Indies, West Africa
Rhynchophorus palmarum (L.)	Stem borer	Central and South America, West Indies
Anacentrinus spp.	Shoots and roots close to the ground	USA
Cosmopolites sordidus (Germ.)	Stem borer	South East Asia, Philippines, Queensland
Diaprepes abbreviatus L. and *D. famelicus* (Oliv.)	Sugar cane root and shoot borer	Central and South America, West Indies

Table 25. Major pests of rice and their distribution
(most important pests marked with an asterisk*)

	Method of attack	Distribution
ORDER: THYSANOPTERA		
Chloethrips oryzae (Williams)	Leaves in nursery	India, Asia, Malaysia, Indonesia, Philippines
Haplothrips oryzae (Mats.)	Leaves and fruit	Japan, Indonesia
H. ganglbaueri Schmutz.	Grain	Malaysia, Indonesia, New Guinea

	Method of attack	Distribution
ORDER: HEMIPTERA		
FAMILY: CICADELLIDAE		
Nephotettix nigropictus (Stål)	Leaf burn, vector of dwarf and yellow dwarf viruses	India, Asia, South East Asia, Pacific
N. cincticeps (Uhl.)		Japan
N. virescens (Dist.)	Vector tungro disease	India, Asia, South East Asia, Pacific
Recilia dorsalis (Motsch.)	Vector rice dwarf virus and orange leaf virus	India, Asia
Cidadulina bipunctella (Mats.)		Philippines
Tetigella spectra (Dist.)		Ceylon, South East Asia, New Caledonia
Thaia subrufa (Motsch.)		India, Asia, Philippines

	Method of attack	Distribution
FAMILY: DELPHACIDAE		
Sogatodes orizicola (Muir)	Vector of Hoja blanca disease	Americas

contd.

Table 25 (*contd.*)

	Method of attack	Distribution
FAMILY: DELPHACIDAE (*contd.*)		
Sogatella furcifera (Horv.)	Vector of rice yellows and stunt disease virus	India, Asia, South East Asia, New Guinea, Australia, Pacific
Sogatodes cubanus (Crawf.)	Vector of Hoja blanca disease	Americas, West Africa
Nilaparvata lugens (Stål)	Vector of grassy stunt	India, Asia, South East Asia, New Guinea, Australia, Pacific
Laodelpha striatella (Fall.)	Vector of rice stripe cereal mosaic and black streaked dwarf virus	China, Japan
Peregrinus maidis Ashm.	Vector of stripe disease of maize and sugar cane	Africa, India, Philippines, Pacific, West Indies
FAMILY: PSEUDOCOCCIDAE		
Ripersia oryzae Green	Leaf sheathes	India, Asia, South East Asia, Cuba, Philippines
Sacchariccus sacchari (Ckll.) (Pink sugar cane mealybug)		Pantropical
FAMILY: COCCIDAE		
Antonina graminis (Mask.)		Americas, Africa, South East Asia, New Guinea, Australia, Pacific
FAMILY: APHIDIDAE		
Rhopalosiphum rufiabdominalis (Sas.)	Rice root aphid	Widespread
Tetraneura nigriabdominalis (Sas.)	Cereal aphid	Pantropical
Schizaphi graminum (Rond.)	Cereals	Widespread

Table 25 (*contd.*)

	Method of attack	Distribution

FAMILY: COREIDAE

Leptocorisa acuta (Thunb.)	Seed heads	Asia, South East Asia, New Guinea, Australia, Pacific
L. oratoria (F.)		Similar distribution but severe in Philippines and Borneo
Leptoglossus australis (F.)		Africa, India, Ceylon. New Guinea

FAMILY: PENTATOMIDAE

Scotinophora spp.	Shoots	Widespread
Nezara viridula (L.)	Stems and grains	Cosmopolitan
Oebalus spp.	Developing seeds	Americas
Thyanta perditor (F.)		Surinam, Venezuela

ORDER: DIPTERA

FAMILY: CECIDOMYIDAE

Pachydiplosis oryzae (Wood Mason)	Shoot deformation	Asia, South East Asia, Africa

FAMILY: ANTHOMYIDAE

Atherigona spp.	Stems	Asia, South East Asia
Hylemya cilicrura (Rond.)		Widespread

contd.

Table 25 (*contd.*)

	Method of attack	Distribution

ORDER: LEPIDOPTERA

FAMILY: PYRALIDAE

Chilo suppressalis (Wlk.)	Larval stem borer	Widespread, Asia
C. polychrysa (Meyr.)		India, Asia, Malaysia, Philippines
C. auricilius Dudg.		India, Asia, Indonesia, New Guinea, Taiwan
Nymphula depunctalis (Guen.)	Paddy case bearer	Africa, India, South East Asia, Australia, South America
Susumia exigua (Butl.)	Leaf worm	South East Asia, Australia, Pacific
Cnaphalocrocis medinalis Guen.	Leaf roller	Madagascar, India, South East Asia, New Guinea, Australia, Pacific
Diatraea saccharalis (F.)	Sugar cane borer	Americas
**Tryporyza incertulas Wlk.	Yellow stem borer	India, Asia, South East Asia, Philippines
*T. innotata (Wlk.)	White stem borer	Indonesia, New Guinea, Australia, Philippines, Malaysia
Rupella albinella (Cran.)		Central and South
(South American white borer)		America
Maliarpha separatella Rag.		Africa, Madagascar, New Guinea

FAMILY: NOCTUIDAE

*Sesamia inferens (Wlk.)		India, South East Asia, New Guinea, Solomon Islands
S. nonagrioides botanephaga T. and B.		Africa

Table 25 (*contd.*)

	Method of attack	Distribution

ORDER: COLEOPTERA

FAMILY: CHRYSOMELIDAE

Dicladispa armigera (Oliv.)	Leaf miner	India, Asia, South East Asia, New Guinea
Asamangulia wakkeri (Zehnt.)	Leaf miner	Indonesia
Rhadinosa parvula (Motsch.)	Leaf miner	Indonesia
R. lebongensis Maulik	Leaf miner	India, China
Trichispa sericea (Guer.)	Leaf miner	Africa, Madagascar
Leptispa spp.	Leaf miner	India, Ceylon

FAMILY: SCARABAEIDAE

Leucopholis spp.	Larvae attack roots	Philippines, Indonesia, Malaysia

FAMILY: CURCULIONIDAE

Lissorhoptrus oryzophilus Kusch. (American water weevil)	Roots and shoots	North and Central America
Helodytes foveolatus (Duval)		South America
Echinocnemus oryzae Mshl.	Roots	India
Tanymecus indicus Faust.	Roots and shoots	India
Hydronomidius molitor Faust		Gujarat

FAMILY: HESPERIIDAE

Parnara spp.	Leaf scaler	Philippines

Table 26. Major pests of coconuts and other palms
(most important pests marked with an asterisk*)

	Method of attack and host	Distribution
ORDER: ISOPTERA		
FAMILY: TERMITIDAE		
Microcerotermes biroi (Desn.)	Nesting in coconut and other trees	New Guinea, Pacific
Odontotermes smeathmani (Full.)	Date palm	Africa
ORDER: HEMIPTERA		
FAMILY: PSEUDOCOCCIDAE		
Pseudococcus adonidum (L.)	Coconut and oil palm	Widespread
Nipaecoccus nipae (Mask.)	Coconut and others	
FAMILY: COCCIDAE		
Aspidiotus destructor Sign.	Coconut scale	Pantropical
Ischnaspis longirostris Sign.	Black thread scale; coconut and palms, fruit	Central and South America, Africa, Indonesia
Pinnaspis buxi (Bch.)	Coconut and other palms, fruit	Americas, Africa, Malaysia, Philippines, New Guinea, Pacific
Chrysomphalus ficus Ashm.		Americas, Africa, India. South East Asia, Australia
Vinsonia stellifera Westw.	Coconut and oil palms	Africa, India, South America
Icerya pattersoni Newst.	Coconut fruits	
Parlatoria blanchardii Targ.	Date palm scale	Cosmopolitan
FAMILY: COREIDAE		
Amblypelta coccphaga China	Nutfall, coconuts	Solomon Islands
Pseudotheraptus wayi Brown	Flowers and fruit	East Africa

Table 26 (*contd.*)

	Method of attack and host	Distribution
FAMILY: TINGIDAE		
Stephanitis typica (Dist.)	Leaves; vector, root wilt disease of coconut	India

ORDER: ORTHOPTERA

Cyrtacantha nigricornis Brun.		South India to Jarva
Pachytylus cinerascens F.		Indonesia, Philippines
Locusta migratoria migratoriodes R. and F.		South East Asia
Valanga nigricornis Krause		Indonesia, Malaysia
Sexava spp.		All coconut countries

ORDER: LEPIDOPTERA
FAMILY: XYLORICTIDAE

Nephantis serinopa Meyr.	Coconut leaf moth	Widespread

FAMILY: ZYGAENIDAE

Artona catoxantha (Hamps.)	Coconut leaf moth	Philippines, Indonesia, Malaysia, New Guinea
Chalconydes catori Jordan	Leaves, oil palms	
Levuana iridescens B.B.	Leaves	Fiji

FAMILY: LIMACODIDAE

Setora nitens (Wlk.)	Leaves	South East Asia
Orthocraspeda trina Moore	Leaves	Sumatra
Darna catenata (Sn.)	Leaves, coconut, sago, oil palms	Celebes

contd.

Table 26 (*contd.*)

	Method of attack and host	Distribution
FAMILY: PYRALIDAE		
Tirathaba rufivena Wlk. (Coconut spathe moth)	Flowers and young fruit	New Guinea, Solomon Islands
T. fructivora Wlk.	Oil palm flowers	Malaysia
Coleoneura trichogramma	Flowers and young coconut fruit	Fiji
Pimeliphila ghesquieri Tams	Oil palm leaf and rachis	Ivory Coast
FAMILY: AMATHUSIIDAE		
Stichophthalma phidippus L.	Leaves	Malaysia
FAMILY: NYMPHALIDAE		
Brassolis sophorae L.	Coconut leaves	Guyana
FAMILY: PSYCHIDAE		
Crematopsyche pendula Joa.		East Malaysia
Mahasena corbetti Tams	Leaves	East Malaysia
Metisa planna Wlk.	Leaves	East Malaysia

ORDER: COLEOPTERA

FAMILY: SCARABAEIDAE		
Oryctes rhinoceros (L.)	Stem borer, palms	India, South East Asia, Pacific, New Guinea
O. monoceros (Oliv.)	Stem borer	Africa
O. boas (F.)	Stem borer	Sudan
Dynastes centaurus (F.)	Stem borer	Sudan
Scapanes australis (Boisd.)	Stem borer (young palms)	New Guinea
Xylotrupes gideon (L.)	Minor stem borer	Widespread South East Asia, Pacific
Leucopholis conephora Burm.	Roots	India

Table 26 (*contd.*)

	Method of attack and host	Distribution
FAMILY: CHRYSOMELIDAE (SUBFAMILY: HISPINAE)		
Brontispa longissima Gestro	Coconut leaf miner	Indonesia, New Guinea, Pacific
Plesispa reichi Chap.	Coconut leaf miner	Malaysia
Promecotheca cumingi Baly	Coconut, oil palm	Malaysia, Philippines
P. papuana Csiki	Coconut oil palm	New Guinea
Coelaenomenodera elaeidis Mlk.	Oil palm and coconut	West Africa
FAMILY: LYMEXYLIDAE		
Melittoma insulare Farm.	Coconut stem borer (larvae)	Seychelles, Madagascar
FAMILY: CERAMBYCIDAE		
Pseudophilus testaceus Gah.	Date palm stem borer; trunk, crown, petioles	Iraq
FAMILY: CURCULIONIDAE		
Rhynchophorus ferrugineus (Oliv.) (Red or Asiatic palm weevil)	Stem borer; coconut, date, sago, oil and other palms	India, Ceylon, Asia, South East Asia, New Guinea, Pacific
R. palmarum (L.) (South American palm weevil)	Coconut and other palms (and sugar cane)	Central and South America
R.phoenicis (F.) (African palm weevil)		Africa
R. bilineatus (Montr.)	Coconut	New Guinea
R. vulneratus Panz.	Coconut	Malaysia
Metamasius hemipterus (L.)	Rotten cane stalk borer—also on bananas and coconut	Central and South America, West Africa
Diocalandra taitiense (Guer.) (Tahitian coconut weevil)		Madagascar, New Guinea, Pacific
Amerrhinus pantherinus Oliv.	Palm leaf stalk borer	Brazil

FURTHER READING

Alibert, H. (1951). Les insectes vivant sur les cacaoyers en Afrique Occidentale. *Mem. Inst. Franc. Afrique noire, Dakar* **15.** 1–174.

Bouland, M. (1967). *Hemipteroides nuisible ou associes aux cacaoyers en Republique Centrafricaine, Premier Partie, Cafe, Cacao, Thee. Paris.* **11.** 220–234.

Cranham, J. E. (1966). Tea pests and their control. *Ann. Rev. Ent.* **11.** 491–514.

Cross, W. H. (1973). Biological control and eradication of boll weevil. *Ann. Rev. Ent.* **18.** 17–46.

Elmer, H. S., Carpenter, J. B. and Klotz, L. J. (1958). Pests and diseases of the date palm. Part 1. Mites, insects and nematodes. *Pl. Prot. Bull. FAO* **16**(5). 77–91.

Evans, J. W. (1952). *The injurious insects of the British Commonwealth,* Commonwealth Institute of Entomology, London, pp. 242

Frölich, G. and Rodewald, W. (1970). *Pests and diseases of tropical crops and their control.* Pergamon Press, Oxford.

Grist, D. H. and Lever, R. J. A. W. (1959). *Pests of rice.* Longmans, London, pp. 590.

IRRI (1967). *The major insect pests of the rice plant.* Johns Hopkins Press, pp. 729.

Jepson, W. F. (1954). *A critical study of the world literature on the lepidopterous stalk borers of tropical graminaceous crops.* Commonwealth Institute of Entomology, London.

Le Pelley, R. H. (1968). *Pests of coffee.* Longmans, London, pp. 590.

Le Pelley, R. H. (1973). Coffee insects. *Ann. Rev. Ent.* **18.** 121–142.

Leston, D. (1970). Entomology of cocoa. *Ann. Rev. Ent.* **15.** 273–294.

Lever, R. J. A. W. (1969). Pests of the coconut palm. *FAO Agric. Studies no. 77,* pp. 190, Rome.

Long, W. H. and Hensley, S. D. (1972) Insect pests of sugar cane. *Ann. Rev. Ent.* **17.** 149–176.

Pemberton, C. E. (1963). Insect pests affecting sugar cane plantations.

Ripper, W. E. and George, L. (1965). *Cotton pests of the Sudan.* Blackwells, Oxford, pp. 345.

Pearson, E. O. and Darling, R. C. M. (1958). *The insect pests of cotton in tropical Africa.* Commonwealth Institute of Entomology, London, pp. 353.

Rivnay, E. (1962). *Field crop pests in the Near East.* Junk, Amsterdam, pp. 450.

Ruinard, J. (1958). Investigations into bionomics, economical importance and possibilities of control of the sugarcane stalk in Java (in Dutch, Eng. summary). Hilversum, pp. 222.

Schmutterer, H. (1969). *Pests of crops in Northeast and Central Africa.* Fisher Verlag, Stuttgart, pp. 296.

Smee, L. (1963). Insect pests of *Theobroma cacao* in the Territory of Papua and New Guinea. *P.N.G. Agricl J.* **16.** 1–19.

Sorauer, P. (1924–1939). *Handbuch der Pflanzenkrankheiten. Bd. 4–6.* Berlin.

Stapley, J. H. and Gayner, F. C. H. (1969). *World crop protection vol. 1, Pests and diseases.* Iliffe, London, pp. 270.

Szent-Ivany, J. J. H. (1961). Pests of *Theobroma cacao* in the Territory of Papua and New Guinea. *P.N.G. Agric. J.* **13.** 127–147.

Szent-Ivany, J. J. H. and Ardley, J. H. (1963). Insects of *Saccharum spp.* in the Territory of Papua and New Guinea. *Proc. Inst. Soc. Sugar Cane Technol.* **59.** 73–79.

Taylor, T. H. C. (1937). *The biological control of an insect in Fiji.* Imperial Institute of Entomology, London, pp. 239.

Tothill, J. D., Taylor, T. C. C. and Paine, R. W. (1930). *The coconut moth in Fiji*. Imperial Bureau of Entomology, London, pp. 269.

Williams, J. R. *et al.* (Ed.) (1969). *Pests of sugar cane*. Elsevier, Amsterdam, pp. 568

Wood, B. J. (1968). *Pests of oil palms in Malaysia and their control*. Kuala Lumpur Inc. Soc. Planters.

Wyniger, R. (1962). Pests of crops in warm climates and their control. *Acta Tropica, Suppl. 7*. Verlag für Recht und Gesellschaft A.G., Basel, pp. 555.

Index